Zukunft und Forschung

Series Editors
Martin Lu Kolbinger, Salzburg University of Applied Sciences, Salzburg, Austria
Elmar Schüll, Salzburg University of Applied Sciences, Salzburg, Austria

The publication series 'Zukunft & Forschung' is of interdisciplinary nature. It contains publications from various fields and disciplines that are characterized by a future-oriented research interest, usually in German language. The series is published by the Research Group 'Society and Innovation' of the Salzburg University of Applied Sciences.

More information about this series at http://www.springer.com/series/8154

Lars Gerhold · Dirk Holtmannspötter ·
Christian Neuhaus · Elmar Schüll ·
Beate Schulz-Montag ·
Karlheinz Steinmüller · Axel Zweck
Editors

Standards of Futures Research

Guidelines for Practice and Evaluation

Editors
See next page

The translation and publication in English language was made possible by the support of

Forschungsforum Öffentliche Sicherheit an der Freien Universität Berlin

nexus Institut

Netzwerk Zukunftsforschung e.V.

Translation by: Lucais Sewell and Dominic Bonfiglio

ISSN 2629-8279 ISSN 2945-7874 (electronic)
Zukunft und Forschung
ISBN 978-3-658-35805-1 ISBN 978-3-658-35806-8 (eBook)
https://doi.org/10.1007/978-3-658-35806-8

© The Editor(s) (if applicable) and The Author(s), under exclusive license to Springer Fachmedien Wiesbaden GmbH, part of Springer Nature 2022
This work is subject to copyright. All rights are solely and exclusively licensed by the Publisher, whether the whole or part of the material is concerned, specifically the rights of translation, reprinting, reuse of illustrations, recitation, broadcasting, reproduction on microfilms or in any other physical way, and transmission or information storage and retrieval, electronic adaptation, computer software, or by similar or dissimilar methodology now known or hereafter developed.
The use of general descriptive names, registered names, trademarks, service marks, etc. in this publication does not imply, even in the absence of a specific statement, that such names are exempt from the relevant protective laws and regulations and therefore free for general use.
The publisher, the authors and the editors are safe to assume that the advice and information in this book are believed to be true and accurate at the date of publication. Neither the publisher nor the authors or the editors give a warranty, expressed or implied, with respect to the material contained herein or for any errors or omissions that may have been made. The publisher remains neutral with regard to jurisdictional claims in published maps and institutional affiliations.

Responsible Editor: Stefanie Eggert
This Springer VS imprint is published by the registered company Springer Fachmedien Wiesbaden GmbH part of Springer Nature.
The registered company address is: Abraham-Lincoln-Str. 46, 65189 Wiesbaden, Germany

Editors
Lars Gerhold
Research Forum on Public Safety
and Security
Freie Universität Berlin
Berlin, Germany

Christian Neuhaus
FUTURESAFFAIRS
Büro für aufgeklärte Zukunftsforschung
Berlin, Germany

Beate Schulz-Montag
Foresightlab
Berlin, Germany

Axel Zweck
Innovations- und Zukunftsforschung
RWTH Aachen University
Aachen, Germany

Dirk Holtmannspötter
VDI Technologiezentrum GmbH
Düsseldorf, Germany

Elmar Schüll
Salzburg University of Applied Sciences
Salzburg, Austria

Karlheinz Steinmüller
Z_punkt GmbH
Berlin, Germany

Contents

Part I Standards of Group 1: The Future as a Subject of Inquiry

1 **Images of the Future** 5
   Christian Neuhaus

2 **Modality** 13
   Karlheinz Steinmüller

3 **Validation by Argumentation** 21
   Armin Grunwald

4 **Aligning the Research with Ambitions for Action** 33
   Gereon Uerz and Christian Neuhaus

5 **Interdisciplinarity** 41
   Elmar Schüll

6 **Transdisciplinarity** 49
   Hans-Liudger Dienel

Part II Standards of Group 2: Good Research Practice

7 **Objectives and Framework Conditions** 61
   Kerstin Cuhls

8 **Transparency** 69
   Elmar Schüll and Lars Gerhold

9 **Theoretical Foundations** 75
   Elmar Schüll

10	Method Selection	83
	Lars Gerhold	
11	Producing Quality Research	93
	Roman Peperhove and Tobias Bernasconi	
12	Scientific Relevance	101
	Birgit Weimert and Axel Zweck	
13	Codes of Conduct and Scientific Integrity	109
	Andreas Weßner and Elmar Schüll	

Part III Standards of Group 3: Practical Relevance and Effectiveness

14	Practical Relevance, Usefulness, and Effectiveness	121
	Edgar Göll	
15	Understanding the Type, Role, and Specificity of the Research Audience	127
	Edgar Göll	
16	Transferability and Communication of Results	133
	Beate Schulz-Montag	
17	Identifying Decision-Making Spaces and Options	141
	Klaus Heinzelbecker	
18	Project and Process Management	147
	Hans-Liudger Dienel	

Introduction

About this Volume

Given the rapid rate of change across all domains of human activity, actors in business, government, and civil society have become ever-more reliant on research into the nature of things to come—that is, on how the *future* may emerge from the complex interaction of current and anticipated trends and relationships. Driven in part by rapid advancements in communications technology, decision-making cycles for executives and policy-makers have become increasingly compressed. At the same time, the range of questions with relevance for strategic decision-making is growing ever larger. What breakthrough technologies are anticipated to emerge? What will be the hallmarks of the workplace of tomorrow? How should one respond to a warming world? These are but a few of the questions currently occupying the attention of thought leaders in various fields.

Thanks to growing recognition for the advantages of using rigorous methods to consider potential future developments and associated opportunities for action, futures studies has made considerable strides in recent years as a discipline. Yet all too often, research findings of questionable quality and provenance compete for the attention of decision-makers and the broader public.

This raises the question: how can one recognize good futures research? That is, what are the hallmarks of a rigorous investigation of the future—one that fulfills scientific standards, does justice to its subject matter, and effectively supports decision-making?

There are no succinct or generally applicable answers to this question. This is attributable to the broad range of activities subsumed under the banner of futures research—including approaches whose theoretical and methodological foundations are less than fully transparent. Among the lay public, for example, it is often assumed that futures researchers are modern-day soothsayers who *predict* the future. This misconception has not augmented the standing of the discipline. Futures research is ill-defined because a great number of academic disciplines and domains of professional practice are necessarily compelled to consider what the future may bring—yet they construct their postulations in the absence of reflection on the standards and guidelines that should inform such activity. The hard sciences, for example, are concerned with developing time-invariant descriptions of deterministic relationships. The focus on causal necessity may occlude the *modality* of the future developments being considered—that is, whether *possible*, *probable*, or *desired* circumstances are at issue. Within narrow disciplinary perspectives, prognostication can quickly succumb to various pitfalls, including the reflexive postulation of continuity (disregarding the potential for change), blind faith in mechanistic models (overlooking important mediating factors), and the refusal to acknowledge uncertainty (ignoring the contingent nature of the future). Alternatively, for ardent empiricists the inherently *contingent* nature of future developments may spawn a refusal to accept the validity of any speculative endeavor.

The need to place futures studies on firmer scientific footing emerges precisely from these points of difficulty. By *scientific* we mean research that is based on transparent, accepted, and rigorous methods, methods that are properly geared to the "future" as a topic of inquiry. Facilitating such research is the aim of this volume, which presents uniform and recognized standards for best practice in futures studies. The standards set forth in this volume have been formulated as guidelines for futures researchers, but they may also serve as criteria for third parties wishing to evaluate particular projects.

Although some degree of consensus on methods has been reached among futures researchers operating in specific institutional settings, this volume represents the first effort to establish a comprehensive set of uniform guidelines for the discipline.

The contributing authors are convinced that for any human undertaking to mature from context-bound considerations to a full-fledged professional discipline, the codification of commonly accepted standards is essential. Indeed, a structured framework for understanding our engagement with the world is essential for experience to be distilled, organized, and transmitted as knowledge.

In line with this insight, one overarching aim of this volume is to gather, consolidate, and extend existing methodological discussions so as to furnish a robust foundation for the expansion and further professionalization of the discipline. The standards presented here do not make a claim to universal applicability, nor are they complete. Rather, they aim to furnish orientation for assessing quality in futures research. In this way, we hope to augment the professional abilities and confidence of future researchers while also bolstering awareness for the important contribution that can be made by futures research to the advancement of knowledge. Indeed, the adoption of these standards by futures researchers promises to considerably enhance the standing and reputation of the discipline.

This guide has been conceptualized for all who are involved in some capacity with research concerning the future, including academics, consultancies, and corporate strategy departments—in short, anyone who hopes to engage with the future as a subject of inquiry in a more rigorous and effective manner. In addition, the volume stands to benefit the organizations that commission futures research, by improving their ability to evaluate project findings.

Preliminary considerations

What exactly is a *standard?* Following Sanders (2006), we view standards as applied principles and methods that are agreed on by experts in a discipline and that contribute to improving that discipline's quality. In accordance with this definition, the standards presented here provide a guide to high-quality work in futures research; they establish the "rules of the game" for conscientious and rigorous research activities. By extension, we also view standards as "statements of intent" addressed to researchers, participants, and stakeholders (DeGEval, 2008, p. 14). Specifically, standards should:

- serve as an instrument of dialog and a professional point of reference for the exchange of ideas on the quality of professional research,
- provide guidance for the planning and implementation of scientific futures studies,
- be a starting point for training and education, and
- furnish transparency for the public assessment of research findings.

At the same time, the standards represent *criteria* for evaluating futures research. When a given project fulfills relevant guidelines, this is a verification of research quality. In this way, the guidelines can serve as assessment criteria for third-party evaluators or for researchers wishing to appraise their own work.

In terms of structure and function, the standards presented here are in line with the recommendations of the Joint Committee on Standards for Educational Evaluation (Sanders, 2006). With regard to content, many of the standards are novel, though we have drawn extensively on pertinent source material from other disciplines, especially in the social sciences.

Guiding Principles and Organization

Guiding principles

A core conviction that informed the writing of this book is that three guiding principles must be pursued in equal measure when conducting futures research: (1) the subject matter under consideration must be suited to the discipline of futures research; (2) the research should be carried out in a scientific manner; and (3) the study design should be effective for achieving its purposes. Each guiding principle poses a particular question: (1) What specific standards result from the fact that futures research explicitly deals with the future as a subject of inquiry? (2) What standards result from the fact that futures research aims to address this subject matter scientifically? (3) What standards result from the special purposes of futures research?

In line with these questions, the standards presented here fall into three groups:

1. The first group of standards all hinge on the defining characteristic of futures studies—namely, the *future* as a subject of inquiry. This group of standards ensures that adequate attention is devoted to the specific output of futures research—that is, the formulation of statements about hypothesized future circumstances. The standards presented under this group pertain to research methods, validation techniques, and approaches for substantiating arguments. They also discuss various needs: to account for the fundamental indeterminacy of the future; to explicitly account for underlying assumptions; and to properly characterize descriptions of the future as possible, probable, or desirable.
2. The second group of standards results from the differences between futures *research* and other approaches for considering the future. This group includes standards that help ensure that assertions about the future are generated in a scientific manner (i.e. according to scientific principles and accepted procedures and techniques). The process, results, and documentation of scientific work are one area of concern. These aspects require, among other things, a clear definition of the research question and careful consideration of the research context. Additional topics addressed under this group include transparency, data quality, theoretical foundations, and method selection.
3. The third group of standards results from the *purposes* of futures research. Accordingly, these standards are designed to ensure that futures research

fulfills its objectives as effectively as possible. Futures research is often characterized by a strong practical focus and by definition is not aimed at the acquisition of temporally invariant knowledge (i.e. laws of physics or biology). The standards presented here ensure a suitable orientation to real-world concerns and contexts. The group includes a large number of application-oriented quality criteria, including in particular criteria with relevance for the management of successful future-oriented research and consulting projects.

Each group of standards is presented with a short introduction that illuminates key considerations. Before drawing on a given standard to undertake a research project or conduct an evaluation, we would encourage the reader to consult the introductions first.

Organization
The chapters are organized as follows:

- *Summary and essentials*: Each chapter opens with a thumbnail description followed by a short discussion to enable quick familiarization.
- *Guidelines*: The guidelines section clarifies how the standards are to be applied while also discussing specific procedural steps.
- *Common shortcomings and pitfalls*: This section describes how the standards may be violated or misapplied, including associated consequences.
- *Illustrative example*: Each chapter presents a real or fictitious case study from futures studies to illustrate the application of the standards and problems that may arise.
- *References*: Space constraints have prevented a full discussion of the relevant literature. The reader is encouraged to consult the sources cited should questions arise.

Following Sanders (2006), the following criteria govern the application of scientific standards:

- The standards aim to guide research activities or their subsequent evaluation, but are not designed as a tool for persistent monitoring.
- Not every standard is applicable to every research project.
- Not all standards can always be fulfilled to the same extent.
- The standards may require modification to suit project-specific requirements.

However, when conducting a project, all of the applicable standards should be taken into account and met to the greatest extent possible.

List of standards

I Standards of Group 1: The Future as a Subject of Inquiry
 1 Images of the Future
 2 Modality
 3 Validation by Argumentation
 4 Aligning Research with Ambitions for Action
 5 Interdisciplinarity
 6 Transdisciplinarity
II Standards of Group 2: Good Research Practice
 7 Objectives and Framework Conditions
 8 Transparency
 9 Theoretical Foundations
 10 Method Selection
 11 Producing Quality Research
 12 Scientific Relevance
 13 Code of Conduct—Scientific Integrity
III Standards of Group 3: Practical Relevance and Effectiveness
 14 Practical Relevance, Usefulness, and Effectiveness
 15 Consideration of Type, Role, and Specificity of the Research Audience
 16 Transferability and Communication of Results
 17 Identifying Decision-Making Spaces and Options
 18 Project and Process Management

References

DeGEval—Gesellschaft für Evaluation e. V. (Ed.). (2008). *Standards für die Evaluation* (4th unchanged edition). https://www.degeval.org/fileadmin/user_upload/Sonstiges/STANDARDS_2008-12.pdf. Accessed 21 June 2021.

Sanders, J. R. (Ed.). (2006). *Handbuch der Evaluationsstandards: Joint Committee on Standards for Educational Evaluation* (3rd revised and updated edition). VS Verlag.

Editors and Contributors

Editors

Univ.-Prof. Dr. Lars Gerhold Research Forum on Public Safety and Security, Freie Universität Berlin, Berlin, Germany
lars.gerhold@fu-berlin.de

Dr. Dirk Holtmannspötter VDI Technologiezentrum GmbH, Düsseldorf, Germany
Holtmannspoetter@vdi.de

Dr. Christian Neuhaus FUTURESAFFAIRS, Büro für aufgeklärte Zukunftsforschung, Berlin, Germany
christian.neuhaus@futuresaffairs.com

FH-Prof. Dr. Elmar Schüll Salzburg University of Applied Sciences, Salzburg, Austria
elmar.schuell@fh-salzburg.ac.at

Beate Schulz-Montag Foresightlab, Berlin, Germany
schulz-montag@foresightlab.de

Dr. Karlheinz Steinmüller Z_punkt GmbH, Berlin, Germany
steinmueller@z-punkt.de

Prof. Dr. Dr. Axel Zweck Innovations- und Zukunftsforschung, RWTH Aachen University, Aachen, Germany
zweck@vdi.de

Contributors

Prof. Dr. Tobias Bernasconi, University of Cologne
tobias.bernasconi@uni-koeln.de

Prof. Dr. Kerstin Cuhls, Fraunhofer Institute for Systems and Innovation Research (ISI)
kerstin.cuhls@isi.fraunhofer.de

Prof. Dr. Hans-Liudger Dienel, Technical University of Berlin
hans-liudger.dienel@tu-berlin.de

Dr. Edgar Göll, Institute for Futures Studies and Technology Assessment
e.goell@izt.de

Prof. Dr. Armin Grunwald, Karlsruher Institute for Technology
armin.grunwald@kit.edu

Dr. Klaus Heinzelbecker, Institut für Sales und Marketing Automation
heinzelbecker@ifsma.de

Roman Peperhove, Freie Universität Berlin
roman.peperhove@fu-berlin.de

Dr. Gereon Uerz, GROPYUS Technologies, Berlin

Dr. Birgit Weimert, Fraunhofer Institute for Technical Trend Analysis in the Natural Sciences (INT)
birgit.weimert@int.fraunhofer.de

Andreas Weßner, Institute for Technology and Work (ITA) e.V.
andreas.wessner@ita-kl.de

Part I
Standards of Group 1: The Future as a Subject of Inquiry

Christian Neuhaus and Karlheinz Steinmüller

The standards presented in the first section of this volume all hinge on the defining characteristic of futures studies – namely, the *future* as a subject of inquiry. This hallmark of the discipline is responsible for various points of divergence from other forms of research.

Futures research involves the systematic and rigorous consideration of potential future circumstances and relationships. However, statements about the future are statements about a special kind of reality. In contrast to disciplines that make statements about past or present reality or about temporally invariant causal mechanisms, futures research advances hypotheses about aspects of reality that have yet to manifest. Indeed, imagined future states are inherently contingent, meaning they may or may not occur. Accordingly, the future may or may not transpire as anticipated, hoped, or feared.

To be sure, future events and circumstances result from the conditions preceding them. However, the influence exerted by these preceding conditions on future reality can never be ascertained conclusively. Hence, while the assertions of futures studies often deal with objects and relationships that exist in some form in the present, these objects and relationships *do not yet* exist in their future form. It is precisely this non-factuality and contingency that makes futures research substantially different from other areas of science.

Clearly, other academic disciplines also construct assertions about conditions temporally removed from the present – such as history, archaeology, or paleontology. But these and other "sciences of the past" can draw on present-day traces of past reality, including the archeological record or the writings of contemporary observers, to substantiate their arguments. Of course, these traces of the past remain in need of interpretation.

This path of validation is closed to futures studies. There are no "traces of the future" in a strict sense. At most, there are present or past conditions with latent potential to influence the future, but whose further development or effect remains fundamentally uncertain. In this way, futures research differs from other disciplines in that the matter at issue *never was, but rather possibly will be*. Accordingly, claims about the future lie beyond direct empirical verification – as long as they are actually claims about the future.

These peculiarities entail a number of special considerations with regard to the research and the proper handling of its findings. Their proper understanding is important for allowing futures studies to make a unique contribution to knowledge production and problem-solving within the sciences.

The particularities of futures studies must be taken into account by consistently regarding posited future conditions as constructed descriptions of a contingent future or, in short, as "images of the future" – and not as predictions of future facts. Indeed, futures researchers must understand that their statements about the future are manufactured images of a possible future, and not descriptions with a claim to factuality (see "Images of the Future"). The formation of such images of the future requires a special type of reflection and guidance, starting with the clarification of the problem at issue. This also leads to one basic tenet of future studies: it must always be made clear that one is talking primarily about the future, and not about the present or the past.

It is also of elementary importance to clarify whether statements about the future are intended to describe possible events in a value-neutral sense, or whether these events represent a future that is desired or feared. Statements in the former category are "descriptive," while those in the latter category are "normative." Whereas descriptive statements about the future provide a *context* for future action, the latter category of statement is meant to impel action for shaping future circumstances (see "Modality").

In both cases, futures studies remain dependent on the current state of knowledge about the present and past. Future circumstances and developments evolve from the reality that precedes them – yet the past does not predetermine the future. The scientifically rigorous generation of statements about the future always requires consideration of the current state of knowledge. In this way, futures researchers must strive to gain the best possible understanding of the specific subject matter they are seeking to address.

As a rule, this requires bringing together perspectives, theories, and knowledge across disciplinary boundaries and real-world contexts (see "Interdisciplinarity" and "Transdisciplinarity"). Cooperating with experts from different academic disciplines as well as individuals from the world of professional practice demands a sensitive engagement not only with disciplinary methodologies and terminology, but also with perspectives that originate from various practical contexts, e. g.

in government, business, or civil society. In this way, researchers must integrate heterogeneous bodies of knowledge and perspectives in light of a common task.

Of course, ex ante conclusions about the future are neither verifiable nor falsifiable in an empirical sense. The assumptions on which such projections are necessarily based can only be postulated and justified by experience. Their *future validity* cannot be proven with certainty, as the conditions of empirical reality are subject to constant change. Any attempt to indisputably validate future prognoses depends on assumptions regarding the continuity of those conditions or causal structures. It is all the more important, therefore, that researchers clarify the assumptions underlying their statements about the future and expose them to critical reflection. In this way, it will be possible to reject conclusions about the future if new empirical findings about the present or past emerge and the original premises must be revised or refuted.

Because statements about the future cannot be empirically verified, their scientific validation requires a special approach. Statements about the future must be interrogated on the basis of their consistency not only with accepted knowledge and theories, but also with informed expectations about the future by other experts and sources. Thus, statements about the future have to prove themselves through argument and a careful examination of the premises on which they are based. Researchers must carefully deliberate on which approaches to use in evaluating the plausibility of their assertions (see "Validation by Argumentation").

Unlike disciplines that study the past, the natural sciences and other studies of (supposedly) invariant subjects, futures research deals with contingent developments that many societal actors seek to shape. Indeed, the fundamental unpredictability of the future is a prerequisite for decision-making and action. Conversely, the findings of futures research can induce actions that will alter the developments originally forecast. In this respect, futures research must take guidance from the ambitions of the funders and the research audience – from the outcomes they seek and the measures they are willing to implement. At the same time, researchers must take into account the consequences, intended or not, that could follow from those ambitions (see "Aligning Research with Ambitions for Action").

Futures research operates precisely at the nexus between a general desire to shape an uncertain future and the desire to find the most suitable actions for doing so in the face of an uncertain future. Understandably, individuals and organizations want as much certainty as possible when planning for the future. But neither meticulous scientific practice nor demands for reliability can unmake the fact that futures researchers produce statements about circumstances that have yet to happen. It is the future's contingency that imbues action and decision-making with meaning in the first place. At the same time, the associated desire for certainty must inherently remain unfulfilled.

Images of the future are a necessary component of decision-making. Without imagining future circumstances and relationships, we would have no way to orient our behavior, and all our actions would be arbitrary. One purpose of futures studies is to improve the evidence base for actions and to provide more reliable ideas about the future. It is important, however, that researchers conscientiously refrain from portraying their images of the future as hard and fast predictions of the world to come.

The standards and guidelines presented in group 1 all address the defining hallmark of futures research: its subject of inquiry is the *future*, or, more precisely, *possible futures*. Together, they provide a methodical and rigorous basis for making statements about something that does not (yet) exist.

Images of the Future

Christian Neuhaus

Summary

Futures researchers understand their work as an attempt to create an image of the future. They base their portrayals on well-founded knowledge about the past and the present. But while future events may be the consequence of previous decisions and actions, it is precisely for that reason that real-life outcomes are so hard to predict. Even the best efforts of futures researchers remain in some sense imaginative: their portraits are not representations of an already determined state of affairs; rather, they are constructions that rely on empirically unverifiable presuppositions in the present to envision a future that has yet to happen. The future, after all, is fundamentally open and contingent. Were it not, we wouldn't endeavor to shape it in the first place. Futures researchers striving for quality emphatically take into account the future's contingency in their approaches and methods. They emphasize it explicitly and unmistakably in discussions with people outside their field by noting that forecasts and scenarios are *constructed images* of the future. Indeed, the quality of a scientifically produced image of the future resides not only in the care and transparency with which it is produced, but also in its refusal to masquerade as a reliable prediction. At the same time, futures researchers make clear that every decision and action with regard to the future necessarily depends on images of what we think could happen or is likely to happen. That is to say, visions of the future influence the decisions and actions we make toward it.

C. Neuhaus (✉)
FUTURESAFFAIRS, Büro für aufgeklärte Zukunftsforschung, Berlin, Germany
e-mail: christian.neuhaus@futuresaffairs.com

Essentials

All academic claims result from certain constructs and presuppositions, but assertions made by futures researchers are conjectural in a very specific way. Their object of study has yet to occur; strictly speaking, therefore, it lies beyond the empirical. In this sense, futures researchers face a special kind of challenge, one that sets them apart from colleagues in other disciplines: namely, how can one assess the academic rigor of claims about contingent developments before they happen?

In order to acknowledge the imaginary character of assertions about the future, there are four key dimensions to keep in mind when assessing the quality of futures research: 1) the *content* of the assertions, including the time horizon; 2) the *normativity* that shapes the image of the future on offer; 3) the *certainty* that the described future developments and events will occur; and 4) the *time-boundness* of the assertions, i.e. the ways an image of the future necessarily reflects the historical moment of its creation.

1. *Content:* Futures researchers study developments in a wide variety of areas and at different points in the future, some closer, some further way. The "content" of an image of the future encompasses both its subject and its position in future time, which is to say, its distance from the present.
2. *Normativity:* The normativity of an image of the future describes the interests, goals, and values that shape it. Specifying this dimension is a question of degree. *Explicitly normative* images overtly state how the future ought to be. Researchers who make such claims do not assert that they necessarily will occur, nor do they rule them out. *Descriptive* images, on the contrary, aim to portray the future as it probably will be or possibly could be (see "Modality"). Nevertheless, descriptive claims are not entirely free of implicit normativity, i.e. the unspoken desires, values and norms that consciously or unconsciously inform them. The normativity of an image should be distinguished from judgments about its inherent desirability, i.e. whether one finds the envisioned future positive or negative.
3. *Certainty:* Every descriptive image of the future is associated with an expectation, subjective or in some cases collective, about whether future events will bear out as predicted. We call this expectation certainty or uncertainty, and, like normativity, it is usually a matter of degree. Images of the future with strong claims to certainty are mainly found in studies concerned with the identification and description of probable futures, as in the case of prognostics (see "Modality"). In images of future developments deemed possible or

merely desirable, certainty plays less of a role, though they may still come to pass.
4. *Time-boundness:* Every image of the future is a product of its time. Understanding the circumstances of its construction requires an awareness of the values and the historical context that shaped it.

Guidelines

1. *Distinctive features characterizing images of the future:* It is important that researchers think about and thematize the properties that fundamentally distinguish assertions about the future from (a) assertions about the present, (b) assertions about unchanging facts, and (c) assertions about the past.
2. *Images of the future are non-factual constructions*: Studies in futures research, whether presented in oral or written form, should emphasize that assertions about the future have a non-factual subject and are speculative in nature. They result from ideas about how the future can or should be.
   – Be sure to stress this quality before presenting assertions about the future.
   – If the risk is high that the assertions about the future will be misunderstood as statements of fact, reformulate them to indicate their speculative nature.
   – The use of "images of the future" (as well as the explicit use of "futures," which differs from the everyday sense of "future" in its singular form) can help underscore the message.
3. *The indispensability of images of the future*: In order to counteract a disappointment of possibly prevailing simple prediction expectations, it is important to communicate that images of the future, even if they lack predictive certainty, are necessary for decisions and actions about the future. Alongside the care and transparency with which they are produced, the quality and utility of these images lie in aspiring to well-founded projections associated with a claim of high certainty, while rejecting the illusion of true prediction (see "Standards of Group 3".)
4. *Projections beyond the empirical:* Futures researchers should understand that empiricism extends only to the present. They should make a clear distinction between the collection and analysis of empirical data, on the one hand, and projections of the future based on empirical data, on the other. Every assertion about the future based on empirical information is a result of a distinct step of construction based on certain assumptions and premises that cannot

be tested ex ante. Researchers must acknowledge that images of the future rest on presuppositional inferences (see "Validation by Argumentation").

5. *Explicitly state the time horizon:* Futures researchers should distinguish the time to which their assertions refer (future) from the time in which the images of the future are created (present). This distinction underlines the fundamental difference between the already realized present and the contingent and shapeable future. Moreover, researchers should explicitly identify the time horizon, i.e. the period in the future that they address.

6. The *time-boundness of the image of the future:* Like all constructions, images of the future emerge at a certain time and historical context, and if not updated, become outdated. Futures researchers should explain the time-boundness of the assertions they use in their work—both their own and those made by others. Hence, futures researchers not only provide a time horizon for the future period they refer to but also indicate the particular period in which their images of the future are produced.

7. *Granularity determined by observers:* The content-related properties of images of the future—level of detail, perspective, thoroughness, complexity—do not belong to the object of study but grow of the constructing decisions in the research process. Deliberate care must therefore be taken to ensure that the properties concord with the research objective. (See "Aligning Research with Ambition for Action" and "Practical Relevance, Usefulness, and Effectiveness.")

8. *Validate without direct empirical evidence:* Because direct empirical validation of descriptive images of the future is impossible, futures researchers must strive to substantiate their assertions in other ways. Specifically, they must carefully review the assumptions, hypotheses, postulates, and arguments that go into their images of the future (see "Validation by Argumentation").

9. *Focus on the research question:* Images of the future should focus on the particular research question and not aim at representing the future holistically or in general. Researchers should explicitly delimit their object of study, i.e. which aspects are considered, including the chosen focal topic, and the wider scope of other related topics. If possible, they should indicate a need for a decision or action as the starting point for the work. State the related problem and research question when presenting assertions about the future (see "Aligning Research with Ambition for Action").

10. *Regard uncertainty as subjective, not objective:* Assertions about the future are uncertain from individual and collective standpoints. Uncertainty is not an objective property of a future reality. Rather, it is a property tied to the image of the future that is the result of a construction, and it can vary based

on the situation and the observer. Moreover, subjective uncertainty depends on individual experiences with the domain of reality under study.

Common Shortcomings and Pitfalls

a) *Failing to make plain that images of the future are constructions*:
Futures researchers may not understand that images of the future are *constructions* or may not sufficiently emphasize it in their work. As a result, images of the future can be misunderstood as true depictions of future developments. This risk is particularly high when addressing those unfamiliar with studies in futures research or other images of the future.

b) *Failing to examine the presuppositions underlying images of the future:*
Researchers may not examine the presuppositions underlying their assertions about future (non-factual) objects. In particular, they may fail to acknowledge that they are *projecting conclusions* about the future based on – but distinct from – current information on the past or on the present. The limits of the applied methodology or the distinctive features and problems of those specific conclusions are not considered.

c) *Failing to identify the time horizon:*
Researchers may fail to state the *time horizon* of the future scenarios under examination, leaving it unclear whether they are addressing a period in the near future or in the distant future, or, even more serious, whether specific assertions refer to the future, the present, or the past. In extreme cases, researchers may give the impression that they are describing states of affairs that remain the same regardless of their time and place.

d) *Failing to state the time in which images of the future were created*:
Researchers may fail to state explicitly the time and historical context in which an image of the future arose, leaving it unclear whether it was created recently or long ago, or whether it was an idea of a particular author or conveyed through media. Likewise, researchers may fail to date sources and other supporting data.

e) *Methodological pomp:*
Futures researchers may try to neutralize or conceal the fundamental limitations and specific characteristics of assertions about the future by making great efforts in their surveys and analysis. Consciously or not, they may use empirical research methods (often with a quantitative emphasis) to create the impression that their images of the future are exact or entirely reliable.

f) *Reducing futures studies to the documentation of current ideas about the future*: A special case of the aforementioned mistake arises when researchers empirically collect or otherwise ascertain *ideas about the future* from other sources — such as experts and stakeholders — and then declare them to be scientific *descriptions of the future* without examining the assumptions and models on which they are based. In this case, the study becomes nothing more than a documentation of current ideas about the future.

g) *Concealing inherent uncertainty:*
Those who commission studies in futures research often want to control the amount of uncertainty associated with the forecasts or scenarios. This is particularly true for contract research meant to solve or alleviate a problem associated with the future. A typical mistake occurs when futures researchers, under pressure from a client, *conceal the inherent uncertainty* of their work. This can also happen when communicating the results of studies to the media. Studies that promise to eliminate uncertainty about the future are popular but they are not scientifically rigorous.

h) *Overemphasizing uncertainty:*
The opposite mistake researchers make is to *overemphasize* the uncertainty of their assertions and the contingency of the subject matter, creating the impression that they lack knowledge and analytical skills. They fail to point out that the uncertainty is not particular to their study but to all assertions about the future, and that the utility of future studies increases when authors grapple with issues related to uncertainty and contingency.

i) *Failing to recognize one's own normative assumptions:*
Images of the future may oscillate between descriptive and prescriptive claims, often unconsciously. Even in the case of strongly prescriptive assertions, researchers can mistakenly believe them to be value-neutral if they *fail to examine their own normative assumptions*. As a result, they present their values, desires, and goals as scientifically proven and universally valid. What is particularly problematic is the intertwining of normative perspectives and attitudes with findings that are otherwise descriptive and academically rigorous. Such an instrumentalization devalues well-founded images of the future and decreases their utility for everyone else (see "Modality").

Illustrative Example

Researchers in an interdisciplinary project are tasked in 2022 with studying the future of X-Burg, a medium-sized city in southwestern Germany. The study is designed to be read by institutions, associations, and other stakeholders in the local government and economy. The results are to be published in spring 2023.

After a thorough analysis of the interests and potentials of local stakeholders and the speed of change of the built environment and the urban economy, researchers set a time horizon of 20 years, so that all projections will be pegged to the year 2042. For most of those involved, the time horizon is meant only to provide basic orientation. It does not represent an exact date, but an approximate period of time in the future distinct from the present. Researchers are careful to note that the prognosis was prepared in 2022.

The project has three phases. In the first, researchers propose and develop six future scenarios for X-Burg in 2042. In the second, they select two to three of the futures that seem most likely under the local policy status quo. In the third, they identify one to two scenarios that are particularly desirable from the standpoint of the local government and economy. The researchers then determine strategic measures for each set of possible, probable, and desirable scenarios, indicating the modality in each case. Wherever possible, they highlight starting points of concerted action for the addressees.

During project preparations and interim meetings with the funding institution, researchers stress that, despite taking the utmost care, their scenarios are created images of the future, not reliable predictions. They describe *possible futures*, not an already determined future that has yet to reveal itself. The researchers present evidence to argue that this is a quality shared by all forecasts and prognoses, regardless of the source. They are also forthcoming about the data used for the scenarios, and about the project's limitations due to its time frame and funding.

While developing the scenarios, researchers determine which issues are pertinent to the project and which are not. For each issue identified, they select models and assessments for ascertaining possible future developments and document their decisions. The objectives and the courses of action available to local decision-makers figure as points of reference in the scenarios. Researchers do not attempt to cover the future of X-Burg exhaustively in the hope, say, of providing a "universal" scenario to please everyone.

Researchers base their conclusions on data collected using a variety of tools from empirical research. They are cautious and aware of the peculiar nature and distinctness of drawing conclusions about the future. They hold long discussions about, say, the usefulness of particular demographic projections from the Federal

Statistical Office or assumptions regarding structural changes that could affect X-Burg. When describing their research or presenting basic information, they distinguish clearly between stating facts and making projections, which go beyond the facts. It is only on the basis of facts and theoretical models that the researchers can paint an image of the city's future.

When the project's intended audience demands to know which future scenario is "correct," researchers point out that they can only say which are most plausible. In doing so, they explicitly state their assumptions and in particular the assumption that the scenarios in question reflect the most likely outcome *in the absence of policy interventions*.

The researchers use this crucial caveat to underscore the basic principle of futures studies: it is impossible to describe the future reality in advance, because the future does not exist and is always contingent and hence indeterminate. It depends on numerous developments — such as gains in knowledge and future decisions and actions taken by actors known and unknown — that have not yet occurred.

At the same time, the researchers stress that, despite the basic contingency of the future, images of the future are indispensable in our decisions and actions toward it. Without them, we would lack orientation. To give direction to our decisions and actions, therefore, we need the best possible images of the future. And the best source for those images is their construction in carefully designed and rigorously conducted futures research.

Further Reading

de Haan, G., & Rücker, T. (2009). *Der Konstruktivismus als Grundlage der Pädagogik*. Lang.
Koselleck, R. (1989). *Vergangene Zukunft: Zur Semantik geschichtlicher Zeiten*. Suhrkamp.
Neuhaus, C. (2006). *Zukunft im Management: Orientierungen für das Management von Ungewissheit in stratregischen Prozessen*. Carl-Auer.
Neuhaus, C. (2009). "Zukunftsbilder in der Organisation," In R. Popp & E. Schüll (Eds.), *Zukunftsforschung und Zukunftsgestaltung*. Springer.

Modality

Karlheinz Steinmüller

Summary

The extent to which an image of the future is possible, probable, and/or desirable is a key concern of futures research. While the first two modalities (*possibility* and *probability*) are explorative, the third (*desirability*) is normative. Good researchers make a clear distinction between these three modalities, not only in each stage of the research project, but also when communicating their findings.

Essentials

Futures research aims to provide a useful orientation for decision-making in the face of an uncertain future. To this end, it addresses three sets of questions:

- What can happen? What can be expected? Based on these questions, possible and probable future developments are explored.
- What do we want to achieve? Based on this question, goals and purposes are identified, normally within the bounds of possibility.
- What can we do? And how can we do it? These questions aim to develop general options for action, including possible strategies and measures. In this regard, action recommendations are crucially informed by expectations concerning future developments.

K. Steinmüller (✉)
Z_punkt GmbH The Foresight Company, Cologne, Germany
e-mail: steinmueller@z-punkt.de

Exploring possible developments and their likelihood of occurrence requires one set of methods; identifying the desirability of such developments (cf. Marien 2002 p. 269 f; Neuhaus 2006, p. 75 f). In this connection, the most crucial point of difference relates to the role of value judgments: While personal bias should be excluded as much as possible from the identification of possible futures, the identification of what is desirable is a fundamentally different task, since it is naturally based on subjective preferences and/or social norms and values. In this way, the fundamental concern of this chapter is to enable researchers to clearly distinguish between *expectations* and *preferences* when designing research or communicating its results.[1]

To be sure, in futures studies (and in the social sciences more generally) absolute objectivity is impossible, as images of the future always depend on a given perspective. Nevertheless, it is a methodological necessity and important on a practical level to clearly distinguish between value judgments about desired or undesired futures, on the one hand, and factual judgments about possible or probable futures, on the other. Otherwise, one may fall victim to wishful thinking (when goals are presented as expectations) or fatalism (when pessimism excessively restricts the possibilities under consideration).

In futures studies, *the modality of possibility* is generally not used to refer to abstract possibilities or events that appear conceivable in principle. In other words, it does not refer to images of the future that merely satisfy the criterion of logical consistency. As a "space of possibility," the future encompasses events and developments that are compatible with the general conditions (economic, technical, societal, ecological conditions, etc.) expected to shape society. The exploration of the future therefore refers to what is concretely possible, to potentialities that can be realized under current or projected conditions. The probability of a specific future actually occurring, however, remains an open question; the key point is that it should not be an impossibility. The identification of possible futures thus involves a tightrope walk between an excessive fixation on the present ("presentism"), which risks overemphasizing current conditions and mediating factors (path dependencies, constraints), and a flight of imagination that risks losing sight of empirical reality.

The modality of probability is grounded in *dominant* expectations, that is, in the sense of justified conjecture about a future occurrence. This future occurrence must necessarily be within the bounds of possibility: for clearly, what is probable

---

[1] We borrow the term "modality" from the domain of logic to denote the characteristics "possible," "probable," and "desirable." However, these terms do not correspond exactly to logical categories such as "possible" and "accidental" or "necessary".

2 Modality

is always first possible. In other words, possibility is a *sine qua non* for probability. A "probable occurrence" is thus a "weak" prognosis that is subject to residual uncertainty. This stands in contrast to certain developments (i.e. "givens") in the development of future scenarios. When presenting various possible futures to audiences, the researcher should categorize them according to their probability of occurrence. If quantifying probabilities is methodologically reasonable, then a future should only be identified as "probable" if its likelihood of occurrence is at least 50% (and is also more likely than all alternatives). To further clarify this point: If a group evaluation has projected the dominant scenario as having a 40% probability of occurrence, it does not represent *the* likely future (only the *most* likely). Within the given portfolio of scenarios, no probable future exists.

The modality of desirability refers to individual and/or collective preferences—i.e. it is an expression of normative values. When considering desirability, researchers' evaluations can range from "to be striven for at all costs" to "to be avoided at all costs." In general, only desirable futures that are also deemed possible fall under the purview of futures studies, for desirable yet impossible futures are unrealistic, utopian wishes. However, in certain cases, consideration of utopian futures may be useful to illuminate values and objectives, or to point the way to a future that is less utopian but also feasible. Researchers must keep in mind that individuals and social groups often have divergent preferences and value systems. The future desired by one group is unlikely to correspond precisely to the future desired by another, given heterogeneous perspectives, interests, and goals. Accordingly, one must give due attention to relevant value systems and interests. One method of doing so is to form a deliberative group tasked with arriving at a desired vision of the future on a consensual basis. Of course, aspects of the future not amenable to modification based on personal or collective preferences should not be considered in such group work.

The above modalities have a bearing on various aspects of the scenarios, including the types that are elaborated, and the specific methodologies used to generate them. Explorative scenarios, for example, are designed to survey the range of possible futures. This may be followed by an evaluation of the scenarios in terms of their probability and desirability. Normative scenarios, by contrast, aim to develop images of the future that are desirable given certain preferences and value systems.

Rigorous research approaches should take all three modalities into account, considering not only possible scenarios and their probability of occurrence, but also their desirability. Achieving clarity about the modality at issue in each stage of a project and about how the stages are interlinked is a basic hallmark of good project design. In this connection, special attention should be devoted to participatory processes, especially when those involved in a project are not yet accustomed to distinguishing between possibility, probability, and desirability.

Researchers should also strive for transparency when communicating their findings—for in the absence of clarification, audiences might confound the desirable with the expected, or the probable with the feared.

Guidelines

1. *Make modality transparent:*
    The most important requirement is transparency—that is, to identify the modality at issue at every step of the research process and when communicating results. There should never be any doubt as to whether one is dealing with possible, probable, or desired futures, neither among the audience of a study, nor among the researchers themselves. Accordingly, researchers should always identify clearly whether a methodology has an exploratory character, and thus serves to reveal possibilities and their probability of occurrence, or whether a normative approach is being used to identify the wishes and interests of stakeholders or the funder.
2. *Use clear and unambiguous language:*
    Futures researchers should use clear, unambiguous phrasing in presentations and descriptions. Modalities should be explicitly named wherever possible. Ambiguity and confusion should be avoided with clear and precise word choice. For example, it may not be clear whether a given vision of the future refers to an objective or merely a possibility.
3. *Avoid implicit value judgments:*
    As a rule, possible futures should be presented in neutral, objective language, because the assessment of desirability should be left to the audience of the study, or at any rate should not be anticipated and imposed. By way of example, in explorative scenarios, authors might inadvertently convey their attitudes in the way they frame descriptions. Explicit and transparent value judgments, by contrast, such as "best case" or "worst case," do not pose a problem, especially if they are developed in a group process with the funder and express the funder's position. In normative scenarios, value judgments are of course necessary, but one should avoid hyperbole, as this can damage credibility.
4. *Make sure that scenarios are plausible:*
    In the case of possible and probable futures, it goes without saying that the scenarios should be internally consistent and at least partially corroborated by existing literature or sources of knowledge (see "Validation by

Argumentation"). Yet desirable futures are also subject to the criterion of plausibility—that is, they should remain within the bounds of the possible (although unanimity on this issue among project participants is not required).
5. *Avoid implicit forecasting:*
Particularly when extrapolating future trends, precise phrasing is essential. Vague or careless word choice may give the unintentional impression of a predictive forecast. It would be misleading, for example, if the phrase "we assume" were to imply that one possible future was an inevitability. Similarly, the framing of the scenario development process should be explicit, to avoid insinuating that possible scenarios left unconsidered have no chance of occurring. When a probable or likely future is at issue, explicit and clear reference must be made to this fact.

Common Shortcomings and Pitfalls

a) *Failing to eliminate problems in the research design:*
The research design should be determined based on the methodological requirements of the project, and not merely based on the methodological experience of the team. The researchers may also fail to adequately consider whether an explorative or normative approach makes sense in each step of the project. The difference between possible, probable, and desirable futures should be kept in mind at all times, and not just in retrospect.
b) *Confounding modalities:*
Researchers may fail to clearly distinguish between possible and desired futures, or fail to identify whether a given future is an expectation or a goal. One common form of this error is jumping too quickly between modalities in a discussion, such that readers assume a probable future is at issue, when the image of the future is merely one possibility. Conversely, when being interviewed, researchers may restrict themselves to discussing probable futures rather than desirable ones, in order to appear more rigorous and scientific.
c) *Self-censoring results:*
Researchers must avoid the temptation to discount possible futures that they have identified in an explorative process merely because they appear unpleasant or contravene prevailing assumptions or norms. Peer pressure may create hurdles to independent inquiry ("That topic is too sensitive"). Similarly, there may be political pressure to ignore unfavorable possibilities ("The client doesn't want to hear that").

d) *Overemphasizing constraints:*
   While the exploration of possible futures must take foreseeable developments into account, one should avoid overemphasizing constraints, whether structural in nature (e.g. existing laws) or related to available resources for action, as this may produce an excessively narrow perspective that ultimately culminates in fatalism.
e) *Failing to screen out implicit value judgments:*
   When describing future scenarios, researchers should avoid injecting personal bias in an intransparent manner. Certain phrasing choices may reveal that the author secretly approves or disapproves of a particular scenario, thus making the project a vehicle for a personal agenda, rather than an exercise in objective inquiry.
f) *Assuming that all stakeholders have identical preferences:*
   When identifying desirable futures, one should avoid assuming that all stakeholders have the same interests, preferences, and values. Differences of opinion and potential conflicts of interest should not be ignored. Researchers should seek to arrive at an explicit consensus, rather than papering over differences with vague language.
g) *Misusing probabilities:*
   The ambiguity of the word "probable" may lead to confusion: When speaking of "probabilities" in the project to refer to the identification of a probable future, one must avoid giving the impression that probabilities will be quantified mathematically. Alternatively, the scenario with the highest probability of occurrence may be misunderstood by readers as a *likely* future, rather than *most likely* among those considered.[2]

Illustrative Example

A few years ago, the German Ministry of Education and Research commissioned a project that aimed to study long-term developments in forestry management and land use. The participating researchers were tasked with considering a variety of trends, norms, and regulatory approaches. The project aimed to further societal debate on forestry policy.

---

[2] Space constraints prevent a more thorough discussion of the numerous mistakes that can occur when calculating probabilities. To name here just one: researchers may erroneously multiply the probabilities of dependent events, generating an excessively low probability estimate.

The project consisted of two clearly defined work packages: the first involved an exploration of possible futures (including scenario generation); the second involved normative questions (specifically, the design of guidelines for desirable futures based on "strong" or "weak" sustainability regimes). A Delphi survey of forestry experts served to explore possible futures (e.g. scenarios for German forests in 2050 and beyond, based on the identified challenges). The survey was also used to illuminate desirable developments (e.g. "What aspects of forestry management should be promoted in the future?"). Due to the clear language of the work packages, there was no danger of conflating modalities.

As the researchers were asked to consider developments up to 2100 (an extreme time horizon, even for future studies), it was necessary to develop scenarios with three time ranges (namely, up to 2020, 2050, and 2100). However, this created a highly complex scenario development process. The consistency analyses ultimately resulted in a total of 14 scenarios—an impractically high number, even considering the divergent time horizons. The participating experts thus suggested combining the scenarios in terms of forestry policy. In this way, normative decisions were taken to enable further exploration.

The original 14 scenarios were distilled into three policy-driven development trajectories (with time ranges up to 2020, 2050, and 2100). The scenarios differed primarily in terms of the level of regulatory intervention that was assumed. All three development trajectories were conceived and constructed as alternative possible futures. The researchers did not conduct a desirability assessment during scenario construction, but rather afterward, based on the scenarios' outcomes and consequences.

When drafting the scenario descriptions, the researchers strived to use factual, neutral, and non-judgmental language, in order to allow readers to judge the scenarios on their own merits. For example, the researchers avoided emotionally charged or tendentious phrases (e.g. "The harvester ripped its way through the forest, belching diesel fumes.") The scenario titles were also modified during the process, due to concern that the titles themselves would engender biased reactions, and thus distract attention away from tangible outcomes.[3]

---

[3] The original scenario titles were "Business as usual," "Strong sustainability policy," and "Free reign of market forces." These titles were later modified to make them more value neutral.

In a subsequent analytical step, the researchers identified the advantages and disadvantages of the three development trajectories, including the forest management policies they implied. In the resulting publication (Z_punkt, 2009), a special section was also included to notify readers that the scenarios represented possible futures, and were intended as a starting point for the discussion of goals and means.

References

Marien, M. (2002). Futures studies in the 21st century: A reality-based view. *Futures, 34*, 261–281.
Neuhaus, C. (2006). *Zukunft im Management: Orientierungen für das Management von Ungewissheiten in strategischen Prozessen.* Carl Auer.
Z_punkt. (Ed.). (2009). *Waldzukünfte 2100: Szenarioreport.* https://z-punkt.de/uploads/files/waldzukuenfte_2100.pdf. Accessed: 22 May 2021.

Validation by Argumentation

Armin Grunwald

Summary

Like any scientific endeavor, futures studies must satisfy criteria of verifiability and validation. However, the usual criteria of scientific verification, such as empirical proof in experiments and measurements or logical deduction from known knowledge, are not applicable to futures studies. Instead, one may apply structural analogies from the *coherence theory of truth* to conceptualize the argumentative validation of knowledge about the future and to underpin it with criteria. It is crucial that the results of futures research are broken down transparently into their constituent parts and that good arguments can be made both for the constituent parts themselves and for the way in which these constituent parts are assembled.

Essentials

The knowledge produced by futures studies—whether founded on model-based energy scenarios, complex Delphi processes, or participatory future workshops—is generated by academic institutes and teams, sometimes in cooperation with non-scientific actors. The claim to a scientific character necessarily implies that these results can be backed up in discourse with good arguments. The root of scientific legitimacy lies in the normative expectation that, in principle, anyone should be convincible by the argumentative tenability of the findings. In case of

---

A. Grunwald (✉)
Institute for Technology Assessment and Systems Analysis (ITAS), Karlsruhe Institute of Technology (KIT), Karlsruhe, Germany
e-mail: armin.grunwald@kit.edu

doubt, the entire chain of argumentation behind the results must be transparently disclosed and critically examined. Openness to critical examination is precisely what distinguishes scientific knowledge from unverifiable "arcane" knowledge.

However, the usual procedures for testing scientific validity are not applicable to the knowledge about the future generated by futures studies:

- There is no possibility of *empirical verification* by experiments and measurements. Statements about future developments or events can neither be verified in reality nor by laboratory observation. Time travel is impossible, and future developments cannot be played out in the laboratory through time lapse procedures.
- In many cases, real empirical testing is replaced by virtual testing using *model-based simulations*. However, the underlying models can only be validated with respect to the past and present, and not with respect to the future. Thus, simulation results can be a component of an argumentative examination but must be relativized under the condition just mentioned and thus come up against fundamental limits.
- Accordingly, the methodological concept of *falsifying* scientific hypotheses as developed by Karl Popper and of approaching "truth" by step-wise corroboration of scientific knowledge against attempts to falsify it is not applicable to knowledge of the future, nor would its application make sense.
- The *logical derivation* of knowledge of the future from knowledge of the present also fails: Even if there were clear laws of progress in social matters, extrapolating them into the future would fundamentally require further premises (e.g. assumptions about the continued stability of this knowledge in the future). But the justification of these premises cannot be decided either empirically or logically (Goodman, 1954; Urban, 1973).

Since the claim to verification cannot be dispensed with unless we are prepared to completely give up on the possibility of generating scientific knowledge about the future, *other validation procedures* are needed. The critical question is what can be covered by an argumentative discourse between proponents and opponents of statements about the future and more broadly, what can and should be defended against doubt.

Knowledge about the future arises through a process that incorporates a broad series of elements, including disciplinary knowledge, causal relationships, model assumptions, value judgments, assessments of relevance and irrelevance, and, in the case of great uncertainty, more or less plausible, partially implicit assumptions. These elements are combined through specific procedures and according to

3 Validation by Argumentation

specific rules to form statements about the future, e.g. by means of modeling and simulation techniques or through participatory procedures. In this way, scientific knowledge about the future emerges from individual components ("ingredients") that are put together ("composed") in a specific way and in a controlled process (Grunwald, 2009). Thus, knowledge about the future is always created at a particular present time and only assessments and knowledge that exist at that moment in time can enter into this knowledge about the future. Since an argumentative examination of knowledge about the future always takes place today, in the present, it cannot take into account whether this knowledge will be correct later, in the future. In view of the impossibility of empirical testing and logical deduction as described above, one can at best make a judgment about the *discursive expectability as of today*. Accordingly, correctness at a later time in the future cannot be made the criterion or standard of argumentative validation (Grunwald, 2009; Knapp, 1978).

Therefore, a dispute about the argumentative value of statements about the future is not about whether the predicted events will occur in a "future present," but rather, the dispute relates to the justification for claims made today based on present knowledge and present assessments of relevance (Knapp, 1978). A discourse on the argumentative reliability of knowledge about the future thus becomes a discourse on the argumentative quality of the ingredients and the composition—that is, the methodological procedure by which the ingredients were "processed" into knowledge about the future (e.g. scenarios). In particular, the argumentative test seeks to uncover the premises, conditions of validity, and assumptions "behind" the ingredients and their composition and to address *their* argumentative tenability. Deconstruction of knowledge about the future into its components as well as the question of the respective premises and boundary conditions reveals that scientific knowledge about the future has a fundamentally *conditional structure* (Acatech, 2012). In terms of argumentative theory, it ultimately consists of if/then chains. It is this decomposability that makes an argumentative examination possible in the first place. This is a precondition for the ideal that, in principle, any person should be capable of forming an opinion about the argumentative tenability of scientific knowledge about the future. Here we find the key to distinguishing scientific knowledge about the future from the "occult knowledge" of soothsayers, clairvoyants, and prophets, and, at the same time, to do justice to the specific qualities of the "future" as an academic discipline.

Guidelines

A variety of established procedures can be used to determine guidelines for an argumentative examination of knowledge about the future. The critical point is that it is always things in the present (theories, assumptions) that are examined. These "present things" are simply projected into the realm of the future. However, due to its essential openness—including the uncertainty of knowledge about the future and the inapplicability of classical procedures for scientific examination—the space of the future has different characteristics, in part, than the spaces of the present or the past. For example, it is a general scientific principle that if you have two mutually incompatible statements about a physical or chemical process, only one can be true. This principle is not always applicable to the future: one cannot rule out the possibility that mutually incompatible, and even completely divergent, statements about the future may prove equally plausible in argumentation (Grunwald, 2013).

One epistemological approach that can deal with this situation is the coherence theory of truth (Rescher, 1973). We will not claim that an argumentative examination can establish the truth of statements about the future. The concept of truth is too open to misunderstanding, and its use would once again suggest the conception already rejected above that the "true" statements about the future are those that prove true later. However, it seems possible and also fruitful to transfer some basic ideas from the coherence theory to the argumentative examination of knowledge about the future to derive guidelines and standards. A transfer of this kind allows for the formulation of guidelines and offers assistance for the argumentative examination of knowledge about the future. These guidelines and rules are directly related to the requirement that knowledge about the future support decision-making and action. According to this epistemological line of thought, the following guidelines must form the basis for an argumentative test of knowledge about the future, and thus also ensure the broadest range of validation in the process of creating this knowledge. The first three guidelines are taken directly from Rescher's coherence theory, while the others are added in keeping with the same line of argumentation:

1. *Consistency:*
   The absence of contradictions is a central requirement for scientific knowledge. Since arbitrary conclusions could be drawn from inconsistent (i.e. self-contradictory) statements about the future, its consistency must be ensured in order to satisfy the expectation that knowledge about the future can successfully guide action. Ensuring consistency is a considerable challenge, especially

3 Validation by Argumentation

in the inter- and transdisciplinary contexts frequently encountered in futures research, since there is usually no common conceptual and theoretical background that could be used as a framework for testing consistency. If the ideal of consistency is most feasible in model-based simulations and scenarios, the problems become more acute in qualitative futures.

2. *Internal coherence:*
   Consistency is a necessary but not a sufficient condition for coherence. The standard of "internal coherence" (Rescher, 1973) refers to the interconnectedness of the individual ingredients. They must be interconnected (e.g. through causal relationships or competitive relationships) in a way that does justice to the presumably complex interactions in the particular field under consideration. At the same time, these interactions must be grounded in robust (of course, present-day) knowledge. Internal coherence thus extends primarily to the (present-day) model building as a basis for obtaining statements about the future.

3. *External coherence:*
   This guideline refers to the interrelatedness of knowledge about the future with parameters of the external world, i.e. the part which has been excluded in a model by the choice of system boundaries. Exclusion does not create a closed system, but requires inclusion of import–export relations at the borders and dependencies on external developments. The argumentative value of knowledge about the future also depends on how well the interfaces of the considered system to the external world are realized and how the interactions are considered. On the one hand, these must be included for the respective present, but on the other hand, possible future changes in these interactions must also be taken into account (based on current assessments).

4. *Adequacy of the system boundaries:*
   The system under consideration and its boundaries must be defined at the outset of a project concerned with generating knowledge about the future, and this definition must be justified, since results obtained later can depend heavily on it. System boundaries are adequate if the knowledge about the future based on them fulfills expectations, e.g. successfully guides decision-making. Accordingly, the ingredients (e.g. knowledge stocks, theories, assumptions about general economic or technological conditions) must be selected in such a way that they contribute to optimally achieving research goals and purposes. Of course, conflicts may arise, and compromises may be necessary if, for example, the ideal requirements for the system to be investigated and the knowledge stocks to be included with regard to argumentative testability cannot be realized for pragmatic reasons (e.g. due to scarce financial or time

resources). Any intentional exclusion of possible ingredients or system components has to be justified argumentatively in the same way as the positive selection.
5. *Epistemological transparency:*
The knowledge stocks used, and especially extrapolations and more or less plausible assumptions, must be scrutinized in an argumentative examination of the underlying premises and their tenability. Thus, futures research should be open and transparent with regard to the epistemological quality of the ingredients, i.e. assumptions and premises should be clearly stated. This applies in general, but especially to those assumptions that are often necessary to close knowledge gaps. Such assumptions are particularly problematic because they are of a hypothetical or speculative nature.
6. *Normative transparency:*
Knowledge about the future cannot be imagined as value-neutral knowledge. For example, normative criteria regarding relevance already permeate the process of defining system boundaries, and ad hoc assumptions may contain normative aspects. Even disciplinary scientific knowledge that is used as a basis for future projections is often not free of normative premises; consider, for example, the differences between neoclassical models of economy and models of evolutionary economics. In addition, there are often explicit normative presuppositions insofar as desired or undesired futures are considered. In all these cases, it is important to disclose the normative premises, values, and interests to ensure argumentative verifiability.
7. *Procedural transparency:*
The argumentative examination must also include the way in which the ingredients have been integrated into (consistent, coherent, etc.) futures. The approaches chosen may include established procedures of modeling, established procedures of aggregating expert assessments (as in the Delphi process), and the observance of established standards of participatory procedures for generating futures.

Common Shortcomings and Pitfalls

Knowledge about the future has to be developed in a much more open space than in the case of knowledge about the past and the present. This means that one-sidedness, short-sightedness, substantive and procedural skewness and other limitations stemming from specific sensitivities (biases) of the respective present

3 Validation by Argumentation

easily give rise to errors. Frequently, these pitfalls are only perceived as mistakes in retrospect. Accordingly, it is far preferable to anticipate such mistakes in order to avoid them.

a) *Adherence to traditional validation principles:*
A common pitfall is to cling to the belief that standard empirical validation is required, whether in the form of empirical proof, or in relation to the need for falsification of a hypothesis. As a consequence, researchers may refrain from hypothesizing about possible future developments (conjectures); disruptive changes are disregarded as "not provable." Ultimately, this constrains the space of future possibilities, such that an excessively narrow range of futures is elaborated.

b) *Striving for "true" knowledge of the future:*
This pitfall is based on the popular misunderstanding that the (scientific) quality of statements about the future depends on whether they later "come true." Statements about the future are then constructed under the condition that they should be "true"—which means in particular that at the end of the analysis, researchers erroneously assume there should only be one unambiguous statement, i.e. the "one correct" forecast, such that all alternatives and uncertainties are neglected or suppressed.

c) *Excessive conservatism:*
Researchers may also be too quick to reject statements about the future as implausible if they contradict widespread ideas or prejudices. In this way, an excessively narrow definition of plausibility requirements may exclude precisely the most interesting and (possibly) the most relevant images of the future. In such cases, the *zeitgeist* is too powerful and leads to "conservative" futures that more or less merely prolong the present. These "conservative" futures are dominant not because the arguments suggest this—they may even contradict the argumentation—but because they conform well to the prevailing conceptions of the world. Especially in such cases, the argumentative examination can serve to reveal imbalances and help to overcome them. These imbalances are the product of the particular present time, and they are quite understandable from a psychological point of view.

d) *Restriction to quantifiable parameters:*
Especially in the model-based creation of knowledge about the future, there is often a preference for quantifiable parameters simply because they are quantifiable and can thus be easily integrated into mathematical models. In this way, qualitative data may be neglected, since they do not "fit" the model and considering them would entail methodological problems. However,

the distinction between quantitative and qualitative data says nothing about the relevance of the corresponding parameters for obtaining argumentatively robust knowledge about the future.

e) *Restriction to available data:*
In a similar way, model-based futures research often prefers to limit itself to parameters for which good and up-to-date data are available. Here, however, the following consideration applies: the availability of data does not guarantee their relevance, nor does non-availability mean non-relevance. Argumentative examination can disclose this difference between relevance and data availability, thus forcing the researcher to reflect on and, if necessary, modify the approach.

f) *Limitation to available experts:*
In all procedures that include expert participation there is a risk that the results will be influenced or distorted by the availability of experts and their willingness to engage. The representativeness of expert panels and the additional participation of non-conforming experts should thus be included as topics of an argumentative examination.

g) *Overconfidence in the model:*
The previous two typical errors are part of a larger complex of frequent misjudgments by modelers. Especially if they have worked for years with certain models and linked part of their identity to them, modelers often tend to overestimate the power of the models and to pay insufficient attention to their premises and presuppositions. When such models are used for projecting the future, this results in argumentatively untenable overestimations of the validity of such simulations.

h) *Naive trend extrapolation:*
In many cases, trends determined empirically or derived from interpretations may be extrapolated into the future without reflecting on the conditions for the admissibility of such extrapolation. In this way, trends are based on past developments and are empirically validated, e.g. by time series. However, it does not follow that these time series can simply be extrapolated into the future. Argumentative examination should ask critically about the conditions under which extrapolation is possible, e.g. whether the conditions and interactions that were instrumental for the existence of the time series in the past may also be plausibly assumed for the future and on which factors and developments this might depend.

i) *Confusing argumentative tenability with the subsequent occurrence of certain future statements:*

Time and again, ridicule and scorn are meted out when certain predictions about the future fail to materialize. Lambasting incorrect predictions is a popular game, especially in the field of economics. However, this ignores the principle cited at the outset that, for methodological reasons, accuracy (in the sense of future occurrence) must not be a criterion for argumentative validity, and that actual falsification ex post does not demonstrate that knowledge about the future was well or poorly justified to the best of one's knowledge and belief at the time it was created (ex ante). Surely, one can learn from what comes to pass or what does not come to pass—but since this learning can only take place ex post, it does not play a part in an argumentative validation.

Illustrative Example

From the 1980s through the early 1990s, there was intensive political discussion about the need for Germany to take a stronger role in space flight. From a technical, economic, and political point of view, the "SÄNGER space transport system" was probably the most ambitious option. The central idea was to develop, under German leadership, a two-stage transport system for manned and unmanned space flights, which—unlike the usual space rockets—could take off and land like an airplane, from European airports, and whose two stages were to be reusable. The lower stage was to operate on normal atmospheric oxygen to reduce the amount of propellant to be carried. While technically ambitious in many respects, particularly with a view to materials engineering and jet turbine design, the main objectives of the project were nevertheless political and economic. The political objective was to enable independent German or European human access to space, and the economic objective was to significantly reduce the exorbitant costs of human space flight. In 1992, a major decision was due to be taken on whether current research into hypersonic technology should be expanded. In order to prepare for this decision scientifically, the German Bundestag commissioned a technology impact assessment (Paschen et al., 1992).

One of the central questions was whether and under what conditions the goal of economic viability could be achieved. Since the development time of the SÄNGER was estimated to be at least 20 years, statements about economic viability had to extend at least 30 years into the future, or better 40 to 50 years. The focus of interest was on two opposing developments: The feature of reusability should lead to a strong reduction of operating costs, e.g. measured in costs per launch or per ton payload, compared to those of disposable rockets. However, the lofty scientific-technical ambitions required that considerably greater expenditures had to be factored into the development phase. The more frequently the space transport system SÄNGER would be used, the more likely the scales would tilt to the side of economic efficiency with respect to both effects. The number of launches per year was therefore taken as the key indicator. Predicting this 20 or 30 years into the future seemed impossible to the project team in view of all the uncertainties (ibid., p. 70 ff.), especially because of the poor predictability of the progress of human spaceflight as a whole and of the competitive situation between the SÄNGER and already existing systems.

For this reason, the future development of space travel was structured in the form of two explorative scenarios: A "conservative" scenario (ibid., p. 72 f.) assumed that space activities would change only insignificantly in scope and manner during the period under consideration. Earth observation and telecommunications would continue to dominate, and the only demand for manned missions would come from the (at that time only planned) International Space Station (ISS).

A "progressive" scenario, on the other hand, assumed a considerably expanded level of activity—for example, from a manned mission to Mars, a "return to the moon" with a manned lunar station for resource extraction, manufacturing facilities in space, or energy generation from solar radiation. In this case, the demand for space activities would be expected to at least double, and perhaps even quintuple (ibid., p. 73). Based on plausibility considerations and ad hoc assumptions, it was deduced that in the conservative scenario, about eight to fifteen launches per year could be expected for the SÄNGER system. Based on model calculations of the expected costs of development and operation based on existing cost models, such a launch rate would be considerably too low to make SÄNGER economically viable, i.e. cheaper to operate than traditional systems (ibid., p. 75). However, in a progressive scenario, any one of the above options would require such a high additional transportation effort into space that the economic viability of SÄNGER would then at least be within reach. It could be shown in this way that SÄNGER would only be economically viable in the progressive scenario. No statement was made about the likelihood of either scenario occurring. However, since in the political arena a progressive scenario was not expected to be realized soon, this conclusion meant the "end" of SÄNGER.

The argumentative examination of the statements about the future was based, on the one hand, on quantitative cost models for the development and operation of complex space transportation systems that were established at the time. The basic features of these cost models were an established standard in the relevant development and planning departments, and they were transparent in terms of modeling and input data, understood in the "immanence of the present" at the time. The scenarios themselves were qualitative in nature. The derivation of expected launch rates for the SÄNGER space transportation system followed rather rough estimates, partly based on ad hoc assumptions that were in principle quite open to attack. Overall, however, the combination of qualitative, primarily narrative framework scenarios with the preceding quantitative model calculations resulted in a sufficiently clear line to provide orientation for decision-making at that time. SÄNGER would only have made sense in the event of a strong expansion of space travel, it was concluded. Whether such an expansion was expectable or made sense was not the subject of this study but was left to political judgment. This example illustrates how, with considerations that are argumentatively contestable in detail, statements can nevertheless be made that were convincing and could provide clear arguments for the decision (that is, the abandonment of the SÄNGER project). In particular, this example shows clearly that the argumentative power did not depend on "pseudo-accuracy" in the details—in other words, on the ostensible accuracy often suggested by quantitative models.

References

Acatech (Deutsche Akademie der Technikwissenschaften, Eds.). (2012). *Technikzukünfte: Vorausdenken – Erstellen – Bewerten. Series Acatech IMPULS.* Springer Vieweg.
Goodman, N. (1954). *Fact, fiction, and forecast.* Athlone Press.
Grunwald, A. (2009). Wovon ist die Zukunftsforschung eine Wissenschaft? In R. Popp & E. Schüll (Eds.), *Zukunftsforschung und Zukunftsgestaltung: Beiträge aus Wissenschaft und Praxis,* 25–35. Springer.
Grunwald, A. (2013). Wissenschaftliche Validität als Qualitätsmerkmal der Zukunftsforschung. *Zeitschrift für Zukunftsforschung, 2* (urn:nbn:de 0009-32-36941).
Knapp, H. G. (1978). *Logik der Prognose.* Karl Alber.
Paschen, H., Coenen, R., Gloede, F., Sardemann, G., & Tangen, H. (1992). *Technikfolgen-Abschätzung zum Raumtransportsystem „SÄNGER".* TAB-Arbeitsbericht, Büro für Technikfolgen-Abschätzung des Deutschen Bundestages. Bonn. https://www.tab-beim-bundestag.de/de/pdf/publikationen/berichte/TAB-Arbeitsbericht-ab014.pdf. Accessed: 28 May 2021.
Rescher, N. (1973). *The coherency of truth.* Oxford University Press.
Urban, P. (1973). *Zur wissenschaftstheoretischen Problematik zeitraumüberwindender Prognosen.* Institut für Wirtschaftspolitik an der Universität Köln.

Further Reading

Habermas, J. (2003). *Truth and Justification.* MIT Press.

Aligning the Research with Ambitions for Action

Gereon Uerz and Christian Neuhaus

Summary

Good futures researchers use various methods for factoring in the possible actions and policy ambitions that are likely to follow on from their work. First, they gather information about the actual practical interests underlying the research question and, importantly, document this information for later reference. Second, they choose an appropriate research design and then construct images of the future whose scope, time horizon, and decision-making parameters are in keeping with the principles of good scientific practice. Third, they carry out a research impact assessment to gauge the potential consequences of the work.

Essentials

Portrayals of the future contribute to the social construction of reality – precisely that aspect of reality most amenable to shaping. As representations of possible future realities, the findings of futures research help to guide present-day decisions about the future. In this way, futures research has the potential to shape the future present.

Futures studies is often linked to an interest on the part of funders or research initiators to mold the future in some way. These ambitions impose various

---

G. Uerz (✉)
GROPYUS Technologies GmbH, Berlin, Germany

C. Neuhaus
FUTURESAFFAIRS, Büro für aufgeklärte Zukunftsforschung, Berlin, Germany
e-mail: christian.neuhaus@futuresaffairs.com

© The Author(s), under exclusive license to Springer Fachmedien Wiesbaden GmbH, part of Springer Nature 2022
L. Gerhold et al. (eds.), *Standards of Futures Research*, Zukunft und Forschung,
https://doi.org/10.1007/978-3-658-35806-8_4

requirements on futures research, especially when designing the study or constructing images of the future. Early on in a project and with reference to the ambition for action, researchers must define the subject of interest, determine the time horizon, identify essential actors, and consider possible options for action.

Explicitly considering possible follow-on actions
Interest in shaping the future (whether on the part of researchers, funders, or third-party addressees) influences a research project's chosen design and methods – and ultimately its results. Good futures researchers make such decisions in a deliberate manner, and align their work with the ambition for action underlying the project. Specific questions help in making decisions about methods, design, and other research questions: (i) What are the specific goals of the research? (ii) Which actors will take action? (iii) How should they take action? For example, whether scenarios can be developed with stakeholders and, if so, which stakeholders should be involved, largely depends on the nature of the action ambitions held by funders or other addressees. In this way, molding specific decisions or action measures may have precedence over an interest in acquiring new knowledge. In such cases, futures studies represents a form of applied science.

Defining the subject area, its determining factors, and its actors
A frequent task of futures researchers is to define the subject area under study in terms of its scope and the factors and actors that shape it. A careful definition of scope is of great importance both for the quality of the research and for the success of the options identified. A scope that is too narrow can neglect important actors or interactions and hence impair the quality of the results and the range of options available. Likewise, a scope that is too large can be too general, making it difficult to identify specific measures. Devoting explicit attention to the underlying ambition to act helps to determine the factors and actors that should be considered when constructing an image of the future.

Determining the time horizon
The determination of the time horizon for research primarily depends on the circumstances and relationships under examination and their specific temporal characteristics. Whether for the short term (t + 5–10 years), medium term (t + 10–15 years), or long term (t + more than 15 years), the time horizon hinges to large extent on: the speed of expected change (e.g. the pace of innovation); the problems being confronted; and on the available resources for action. Other factors that require consideration include the speed with which actors desire change, and the time window for necessary actions. This also includes the question of

whether it is *too early* or *too late* to initiate a futures studies project. Once again, the project design and ambition for action are the main indicators for the choice of time horizon.

Determining options for action
The greater the ambition for action, the greater the need to identify options and estimate their future effects (see the group 3 standards). Indeed, servicing these ambitions is where futures research set itself apart from other disciplines. The inherent openness of the future and the significant degree of uncertainty it brings requires a special type of approach. Of course, it is impossible to determine in advance the best measure or policy with absolute conviction. Futures studies has the unique task of presenting a spectrum of options for action and highlighting the differences between the possible consequences in each case. This can be done both within the same envisioned future situation or in the form of future situations that have changed as a result of the respective options for action (impact scenarios). In this regard, the consideration of specific desires to affect change and the exploration of possible action alternatives are not only *not* mutually exclusive; they are perfectly complementary.

Thinking about intended consequences and possible knock-on effects
The direct relevance of futures studies projects for decisions and action often means contending with potentially far-reaching consequences. Scenarios and other images of the future are designed to guide perceptions, decisions, and actions vis-à-vis the future; as such, they bring with them consequences for reality both intended and not. In light of both types of effects, researchers would be well advised to assess the potential impacts of their work. To be sure, *ex ante* estimation of the potential consequences of a project requires a considerable amount of reflection and effort. However, an appreciation on the part of researchers for how they contribute to the social construction of reality and to the measures taken by real-world actors can increase the value of their contribution. Such critical and self-reflective practices could perhaps be labelled "second-order futures studies."

Guidelines

1. *Clarify and document the ambition for action*: When defining project objectives and tasks, researchers should clarify and document the underlying ambition for action and desire to affect change so this can serve as a point of reference during their work. Researchers should play an active role in the process and

voice any criticisms they might have (e.g. that the range of considered effects of action is too narrow).

2. *Define the project scope in accordance with the objectives for action*: Researchers should clearly define the scope of the project in accordance with the identified ambition for action and desire to affect change, and possibly delineate it from related research fields. They should define the scope so that that decisions and formative interventions covering the intended effects can be made on the basis of the research results.

3. *Identify the central factors and actors*: Researchers should identify the determining factors and actors that could significantly affect the development of the domain in question within the chosen time horizon. A careful mapping of these elements and their interdependencies in the form of a network analysis is indispensable. The analysis of past and present forces in the domain in question should be supplemented by the identification of new actors that could play a significant role within the set time horizon.

4. *Define the time horizon in accordance with the ambition for action:* The time horizon for forecasts and future scenarios should be chosen based on the ambition for action and desire to affect change, and this horizon should be kept in mind during the research project. Researchers must weigh the dynamic properties of the research problem in relation to action-related resource and time constraints, given the specific capabilities of futures research to generate insight.

5. *Reappraise research decisions:* The subject area, mediating factors, relevant actors, and the time horizon are guideposts for the research work, and hence must be determined early on. Nevertheless, researchers need to reflect continually on these aspects and adjust them if the results do not accord with the identified ambition to affect change. This might be necessary if, for example, researchers develop a better understanding of the subject area during their work, or if ambitions change due to interim findings of the research project.

6. *Present alternative courses of action:* Researchers should explore alternative courses of action in terms of their possible future (positive and negative) effects in the subject area and action domain. In doing so, they should make direct reference to the ambition for action identified at the project's outset. If possible, researchers should present a spectrum of options, whose effects will vary within and between scenarios, and which will be associated with different levels of uncertainty.

7. *Reflect on the consequences of the research:* Good futures researchers should consider the potential consequences of their work ex ante as thoroughly as possible. On the one hand, they must consider the ambition for action of the

funder and other addressees along with the intended and unintended consequences that may directly arise from later action measures. On the other hand, they must think about the indirect consequences of their research, i.e. the ways in which the images of the future they create can, once in circulation, shape the social construction of reality and reality itself. Then, they must decide whether to endorse the possible effects of their work, to try and improve those effects, or to scrap the study.

Common Shortcomings and Pitfalls

a) *Failing to clarify the research objectives:* Researchers may fail to discuss in detail the ambition for change held by the funder or client. As a result, the impetus for the research question, and possibly the question itself, may remain unclear. In this way, the project may proceed in an arbitrary fashion, characterized more by chance than by reason.

b) *Failing to contain bias:* This pitfall occurs when the researchers are led by their own interests that do not coincide with those of the client. The research is thus tailored to convince the funder of the researchers' agenda.

c) *Selecting an inappropriate scope and/or time horizon:* (i) Researchers may choose a project scope that is too broad or too narrow and/or (ii) a time horizon that is too long or too short. Neither the research funder nor the researchers provide a clear definition of the domain of interest and/or agree on a time frame that is appropriate to the research question. For example, they may underestimate or overestimate the speed of change within a given field.

d) *Providing a restricted selection of options for action:* Researchers may fail to sufficiently explore the range of possible decisions and options. This is often the case when the funder is interested only in external, retroactive justifications of previous decisions. Alternative options are not explored or considered.

e) *Failing to think about the possible consequences of the research:* Researchers may fail to consider the possible indirect and knock-on effects of a futures study, including in particular the intended and unintended consequences of the future actions. This may be due to a lack of due diligence or to a deliberate decision because of time or funding constraints. Alternatively, it may be motivated by a desire to insulate the funder from uncomfortable facts, or other normative or political considerations.

Illustrative Example

In 2021, a leading global manufacturer in the industrial engineering sector hires a consulting firm to develop a series of future scenarios. The manufacturer wants to consider "disruptive" scenarios as well as a scenario of the most likely developments in the industry and its related sectors (the "base case"). The general objective is to help the company over the medium and long term (i) to become more aware of the larger economic environment and (ii) to prepare for the possibility of radical changes within it. The scenarios are also meant to provide a basis for a strategic yearly discussion between the company's executive board and its clients.

In the short term, the project is intended to create a critical review of the company's research and development priorities and the projected future customer behavior and requirements. Moreover, the scenarios will be used to identify gaps in the company's service and technology portfolio. The overall goal of the project is to examine the company's strategic orientation in view of possible changes in its larger economic environment and to identify implications for future decisions (regarding, say, new products and business models).

The consulting firm discusses the project objectives directly with the company's CEO. It emerges that the project is to be attended by a reshaping of the corporate culture, with a special emphasis on risk awareness. Researchers conduct individual interviews with the senior management teams to assess the company's current situation, its central challenges, and the most important players and other relevant factors in the sector. Due to the long investment and innovation cycles in the industry, the time horizon is set to 2045. The researchers clarify and document the complex project goals and the measures the company intends to carry out at an early stage.

After the desk research and expert interviews, researchers describe the key trends and drivers for the scenarios, discuss them with senior management, and then perform subsequent fine-tuning. They organize a workshop with the management to analyze the interrelationships between various trends and drivers. The consulting firm identifies the mutually exclusive possible future states of factors relevant to the scenarios, discusses them with the management, and consolidates them. The development of the scenarios is software-supported and results in four with a disruptive character and one as a base case. All scenarios are discussed with the senior management in terms of their short- and medium-term consequences and implications for action, especially regarding possible extensions of the value chain (products, services, forward and backward integration). The

4 Aligning the Research with Ambitions for Action

review of the portfolio and, if necessary, its expansion are the main measures under consideration.

Neither the number nor the radical nature of the scenarios pose a problem for the company, as the focus was chosen at the instigation of its CEO. The scenarios and the key indicators for the early identification of changes in the larger economic environment lead to an intensified examination of the company's business environment and to investment decisions in new products and services. The careful definition of the research field and the thorough identification of important factors and actors make it easier to define options for action. The scenarios serve as a central instrument in strategic discussions with customers. Such discussions allow anticipated future developments to be considered in relation to potential action measures by the company and its customers.

Interdisciplinarity

Elmar Schüll

Summary

A systematic exploration of possible future events requires understanding a broad spectrum of factors—at once social, economic, technological, ecological, and political. Outstanding work in futures studies, therefore, almost always brings together multiple disciplinary perspectives and bodies of knowledge. But therein also lies one of the field's fundamental challenges.

Essentials

Academia today consists of disciplines whose boundaries are defined by specific problems, subject areas, theoretical assumptions, and interests. For example, sociology focuses on the social world and relies on theoretical frameworks that address the motives and structures of social interaction. It is this focus, and the ignoring of everything non-sociological, that constitutes the core of sociology. The separation of research into ever more specialized disciplines reflects the increasing division of labor characteristic of modern societies more generally.

To be sure, the organization of academia into disciplines has furthered intellectual progress, but this division is to some extent artificial. The world around us does not follow disciplinary boundaries, and the gap between specific knowledge gains and real-world problems is growing ever wider. More importantly for our purposes, a single-discipline approach is too restrictive for exploring future

E. Schüll (✉)
Salzburg University of Applied Sciences, Salzburg, Austria
e-mail: elmar.schuell@fh-salzburg.ac.at

events. It is impossible to know ex ante which aspects and factors will be crucial for a particular future development. Taking an interdisciplinary approach is the only sensible way to approach an indeterminate future.

Interdisciplinarity does not mean jettisoning disciplinary frameworks and knowledge stocks or calling into question the utility of specialization. Rather, it surmounts disciplinary boundaries for the sake of a particular research question. Indeed, by combining the perspectives, theories, and approaches of different fields, reseachers stand to achieve better insights than would have been possible from within the confines of a single discipline.

Researchers themselves must decide which disciplines to draw on in a specific project and how to organize their collaboration. There is no universal rule stipulating which combination of disciplines produces the best results for a given research question (see Kaufmann, 1987, p. 66). The following guidelines are nevertheless designed to help researchers with their decision-making:

Guidelines

Interdisciplinarity in futures studies does not arise on its own, and it is not achieved by simply cobbling together researchers from different disciplinary backgrounds. Compared with work in a single discipline, successful interdisciplinary research requires more preparation and the fulfilment of more organizational prerequisites:

1. *Take into account disciplinary differences*:
   Collaboration between different subjects in a single discipline or in neighboring disciplines (such as sociology and political science) is easier than that between disciplines that are far apart (such as engineering and ethnology). For projects involving very different disciplines, therefore, special attention must be paid to coordination and oversight, as well as to communication between team members.
2. *Choose leaders with interdisciplinary experience*:
   Interdisciplinary projects depend on having researchers from suitable fields and on a useful framing of the research problem. Leaders with experience in multiple fields and disciplines (see Kaufmann, 1987, p. 72) or in leading interdisciplinary teams can help ensure that a project has a strong foundation.
3. *Assemble a competent and cooperative team*:
   Research team members should have expertise in their respective field and be open to insights and approaches from other disciplines.

4. *Reflect on one's own discipline*:
   The disciplinary background of each research team member brings its own preferences regarding methods, models, concentrations, and the like. It is important that researchers reflect on their own disciplinary preferences and explain them to their team.
5. *Define the research question*:
   Leaders should precisely define a common research question for the group and make sure researchers stick to it. By no later than the start of the research proper, though preferably beforehand, leaders should involve the team in a careful discussion of the disciplinary and interdisciplinary issues that might arise.
6. *Build a consensus*:
   Those involved in the research process must set common objectives, agree on common perspectives and terminology, and develop a common theoretical framework. The different approaches must be integrated into something new that is nevertheless shared by all, though researchers may regard it as a comprise that deviates from their preferred views, models, and methods (see Defila et al., 2006, p. 35). A common basis for interdisciplinary collaboration must also be found with regard to discipline-specific behaviors (think "discussion cultures," say), metatheoretical positions ("What counts as a viable explanation?"), and publishing obligations.
7. *Ensure regular communication*:
   In order to avoid misunderstandings, teams should provide each other with background information about the topics in group discussions before they take place. Repeating background information during a discussion or across several discussions is permissible and in some cases even necessary for the purpose of building understanding (see Immelmann, 1987, p. 87).
8. *Integrate results from each discipline*:
   It is important that results from different disciplines be combined into a coherent whole from the outset of the project (see Defila et al., 2006, p. 36). Frequent discussions about interim results and subgoals—through the establishment of common milestones or separate work packages—help connect differing vantage points.
9. *Disseminating research results*:
   Generally, the addressees of futures research do not belong to a single discipline and often are not even academics or scientists. Hence, typical outputs for disseminating research such as conferences or journals are not useful. Instead, results must be put in a form that the target audience can understand and utilize (see "Transferability and Communication of Results").

Common Shortcomings and Pitfalls

a) *Raising false or exaggerated expectations*
   - *Devaluing disciplinary approaches*: Sometimes the call for interdisciplinarity is accompanied by the devaluation of disciplinary findings. This fails to recognize that interdisciplinarity always presupposes disciplinarity. Generalists who casually discount specialized knowledge implicitly reject a system whose progress has only been possible thanks to centuries of increasing specialization.
   - *Overvaluing holisticism*: It is often assumed that interdisciplinarity has the goal of understanding a subject holistically. Researchers in futures studies sometimes believe that the holistic promise of interdisciplinarity can overcome the future's indeterminacy. In reality, interdisciplinary research can at best produce new insights from integrating different perspectives. Its insights are not holistic; they, too, are partial and incomplete. But they are also more comprehensive than the insights produced by a single discipline and ideally include those aspects that the target audience finds crucial (Heckhausen, 1987, p. 138).
b) *Failing to understand the process*[1]
   - *Falling short in terms of theory and method*: The challenges of interdisciplinary research are underestimated when researchers assume that, in contrast to conventional disciplinary research, no special precautions need be taken. Those involved often know too little about which processes can best build consensus and integrate perspectives. Alternatively, they are unable to apply them correctly, producing a study that is unreflective, arbitrary, and haphazard. Synergistic potentials may go unused and the results may be no different from those that could have been attained from a single discipline.
   - *Failing to monitor group dynamics*: Interdisciplinary projects usually require teamwork over a longer period, which may be new and unfamiliar for some researchers. Often, team leaders underestimate or overlook group dynamics. A possible consequence is that they fail to support communication and either ignore or fail to resolve conflicts.
   - *Failing to monitor quality and provide incentives*: In the academic world, quality control usually takes place within disciplinary boundaries. As a result, work that other disciplines might regard as sloppy is less likely

---

[1] For the following, see Defila, Di Giulio, Scheuermann 2006, pp. 42–44 – unless otherwise indicated.

to harm the reputation of its authors. Conversely, compared with work in a single discipline, interdisciplinary research is less likely to *improve* the researcher's reputation and career chances, and as a result tends to inspire lower levels of engagement (Kaufmann, 1987, p. 78).

c) *Failing to surmount communicational difficulties between disciplinary cultures*
– *Failing to prevent misunderstandings and terminological disputes*: Researchers in a particular discipline tend to take its knowledge and terms for granted. Interdisciplinary work often confronts them with researchers who either are ignorant of these implicit assumptions or do not acknowledge their validity. If a discipline's knowledge and theoretical framework are not sufficiently clear, its expertise will fall on deaf ears among the uninitiated. Conflicts can arise when researchers fail to reflect on their own vantage points and insist that their views are the only ones that are correct or meaningful. With interdisciplinary projects, it is especially important that researchers arrive at definitions together and carefully describe their scope and validity with an eye toward their relevance for the final outcome. In this way, they can forestall conflicts arising from misinterpretations.
– *Pursuing different subjects and questions*: Each discipline has its own perspective on which aspects are relevant to the research question and on how to describe them. If a team is to develop a shared view of the research objective, it is crucial that they work to overcome the disciplinary myopia of its members.
– *Applying different methods and criteria*: An interdisciplinary project relies on the methods and criteria of each of the disciplines that constitute it. The problem is that ideas about which methods and criteria are most appropriate can differ from one discipline to another. It is thus important that the project's research design and methods be clarified at the outset (see "Method Selection").
– *Failing to eliminate prejudices*: Researchers often possess misconceptions about other disciplines. If left uncorrected, they can lead to prejudice and erroneous expectations about what researchers can contribute to a project. Conflicts can arise when researchers claim to know another discipline better than its own practitioners while dismissing its value.

Illustrative Example

The government of a prosperous German state wants to gauge the challenges that the region's employers, employees, and public institutions will face over the next 20 years in response to demographic change. After announcing a call for research proposals, state officials award a three-year contract to a university in the state capital.

The research director is a professor of political science with a Ph.D. in economics who has coordinated several applied research projects at the university. The other core team members consists of a Ph.D. in sociology and a demographer who will be employed as a research assistant at the university for the duration of the project. The milestone deadlines are negotiated over several rounds. The state government insists that the results be made available at least nine months before the end of the legislative session so that they do not interfere with the next election campaign. This deadline was added subsequently to the original project proposal.

As the project begins, its leaders schedule numerous meetings to develop a common understanding of the subject area and to discuss a variety of theoretical models relevant to the research question. Though the leaders have prior experience, they underestimate the effort involved. So they divert energy from other work areas and continue discussions until they come to a common understanding.

The leaders commission several expert reports from researchers in disciplines relevant to the project, including labor and constitutional law, the sociology of work, and business administration. The reports prove very helpful, providing a basis for discussion and additional sources of information. In addition, they conduct interviews and workshops with pertinent stakeholders such as employers' associations, health insurance funds, employee representative bodies, and the Federal Employment Agency.

The project researchers document each step of their work and the head encourages them to publish the most important interim results in journals from contributing disciplines.

While collecting the data, the researchers note the significant differences between disciplinary ideas and interests. To put them in perspective, they weigh the differing viewpoints in concert. The research team then decides to define the terms most relevant to the topic in a glossary. This task was not included in the original project proposal and no time resources were allocated to it. However, the project head convinces state officials of its benefits and they agree to a new timeline.

Officials express interest in tangible and descriptive results that serve as a starting point for a range of policy options. The project head had already assumed as much and included scenarios in the research design. But the team members differ on which type of scenario is best. The demographer argues for a "calculated" scenario for population development such as a parameter constellation. The director and the sociologist insist on a detailed sociological analysis. They regard quantification as problematic and believe it to be dispensable, at least for the time beginning.

The research team sets about preparing and processing the collected data in line with the project's general objective. As part of the scenario preparation, they hold several workshops to discuss the factors governing how demographic change will affect the region. The researchers work well together despite their differing backgrounds because the research question is clear and mutually agreed on. Ultimately, they arrive at very similar assessments. The representatives of the employers' and employees' associations, the business owners, the officials from the state ministries, the Federal Employment Agency, and the workshop participants are happy with the meetings. They feel treated as equal partners and are well informed about the research. They think regularly about the issues related to the project and tell colleagues about their participation.

Evaluating the workshops proves more time intensive than expected. The researchers had expected at least some rudimentary generalizations, but the workshop reports were mostly about individual experiences. While the research team finds empirical data on many of factors identified in the workshops, it lacks specific expertise to interpret some of the information. So the university invites faculty from relevant departments to participate in the discussions.

After interpreting the workshop reports, the research team formulates four scenarios for future demographic developments in the state. The scenarios describe the main policy options and their effects on the regional labor market. The scenarios also highlight government options for supporting companies and employees in the region.

References

Defila, R., Di Giulio, A., & Scheuermann, M. (2006). *Forschungsverbundmanagement: Handbuch für die Gestaltung inter- und transdisziplinärer Projekte*. vdf Hochschulverlag.
Heckhausen, H. (1987). Interdisziplinäre Forschung zwischen Intra-, Multi- und Chimären-Disziplinarität. In J. Kocka (Ed.), *Interdisziplinarität: Praxis – Herausforderung – Ideologie*, 129–145. Suhrkamp.

Kaufmann, F.-X. (1987): Interdisziplinäre Wissenschaftspraxis. Erfahrungen und Kriterien. In J. Kocka (Ed.), *Interdisziplinarität: Praxis – Herausforderung – Ideologie*, 63–81. Suhrkamp.

Immelmann, K. (1987): Interdisziplinarität zwischen Natur- und Geisteswissenschaften – Praxis und Utopie. In J. Kocka (Ed.), *Interdisziplinarität: Praxis – Herausforderung – Ideologie*, 82–91. Suhrkamp.

Further Reading

Defila, R., & Di Giulio, A. (1998). Interdisziplinarität und Disziplinarität. In J. H. Olbertz (Ed.), *Zwischen den Fächern über den Dingen? Universalisierung versus Spezialisierung akademischer Bildung*, 111–137. Leske + Budrich.

Olbertz, J. H. (Ed.). (1998). *Zwischen den Fächern über den Dingen? Universalisierung versus Spezialisierung akademischer Bildung.* Leske + Budrich.

Transdisciplinarity

Hans-Liudger Dienel

Summary

The questions pursued in futures studies tend to be highly complex. As a result, they often necessitate the integration of knowledge from academia and from professional practice. To this end, academic researchers are regularly required to enter into a productive collaboration with experienced partners from the domains of business, government, and civil society. As an undertaking, transdisciplinary research thus has a wider range than both *multidisciplinary* research (which involves academic disciplines working *alongside* each other) and *interdisciplinary* research (which involves *integrating* perspectives from various disciplines). The collaboration that is required in transdisciplinary research demands a significant investment of time and resources. However, the potential gains are also significant, as the transdisciplinary project promises to generate highly innovate findings that are also robust, given their academic rigor and real-world orientation. *Participatory* futures research can be viewed as inherently transdisciplinary, as it involves enriching academic expertise with input from various stakeholders, including the broader population.

H.-L. Dienel (✉)
Department for Work, Technology and Participation, Technische Universität Berlin and Nexus Institut for Cooperation Management, Berlin, Germany
e-mail: Hans-Liudger.Dienel@tu-berlin.de; dienel@nexusinstitut.de

© The Author(s), under exclusive license to Springer Fachmedien Wiesbaden GmbH, part of Springer Nature 2022
L. Gerhold et al. (eds.), *Standards of Futures Research*, Zukunft und Forschung, https://doi.org/10.1007/978-3-658-35806-8_6

Essentials

The key hallmark of transdisciplinary research is its integration of knowledge from academia and professional practice. In this connection, the goal of expanding scientific knowledge is generally subordinated to the practical concern of solving a socially relevant problem. Hence, research activities are generally organized and planned according to practical realities—and not primarily according to scientific considerations. The involvement of practitioners and experts with hands-on experience usually generates findings with a higher practical value, while also improving scientific knowledge gain. At the same time, the involvement of academics in a project with a high level of practical relevance also generally improves the quality and applicability of the project's findings. In contrast to *interdisciplinary* research, where boundaries between disciplines often blur or disappear, *transdisciplinary* research cultivates and exploits the divergent perspectives offered by individuals from various academic and professional settings.

However, the collaboration between academics and professionals can often pose problems. For example, university researchers, particularly in the social sciences, may "look down on" practicing professionals, pigeon-holing them as interviewees or research subjects. By contrast, academics in the hard sciences, such as engineers or chemists, typically have an easier time cooperating with members of the business community. While this may be attributable in some cases to past entrepenurial experience, the "mental distance" between engineers and business professionals is also smaller. This may explain why projects with a practical business concern rarely avail themselves of insights that could be provided by social scientists.

While disciplinary identities often blur or ultimately evaporate in *interdisciplinary* projects, in *transdisciplinary* projects, the self-identification of individuals with their discipline may increase over the course of the project. While this tendency has benefits, due to the cross-fertilization that can occur from divergent perspectives, for a transdisciplinary project to yield valuable results, it is essential for the project participants to be trusting, willing to learn, and committed to open communication. Accordingly, good interpersonal collaboration and effective project management (see "Project and Process Management") are particularly important in transdisciplinary projects.

Futures research relies on transdisciplinary cooperation, as it frequently aims to furnish specific recommendations for real-world action. For this reason, researchers must sufficiently take into account the practical knowledge and perspectives of funders, audiences, and other stakeholder groups (see "Understanding

6 Transdisciplinarity

the Type, Role, and Specificity of the Research Audience"). Furthermore, the precise research methods employed must be considered and defined separately in each project.

The scenarios that are elaborated in futures research often integrate quite divergent academic perspectives, while also addressing practical problems and issues in the domains of business, politics, and society.

Future studies itself is not a standard discipline, but rather an "interdiscipline" that draws on methods and themes from many disciplines and real-world settings. For this reason, futures studies lack a clear disciplinary standpoint. While this can facilitate the reconciliation of frictions between disciplines, it may also weaken the ability of futures researchers to appreciate the typical viewpoints and objections that are native to other fields of knowledge. For this reason, when futures researchers spearhead transdisciplinary projects, there is a special need to remain sensitive to how other academics and professionals "tick." Indeed, for successful transdisciplinary work, researchers should seek to foster an environment of mutual trust, empathy, and open communication.

The 1950s were an important decade for the development of futures studies. In particular, the fields of cybernetics and mathematical modeling encouraged positivist and prognostic orientations. In order to integrate real-world expertise, Delphi analysis was developed, which remains an important method to this day. With the growing recognition that transdisciplinary cooperation was a necessary component of research, techniques were developed for collaborating with professionals outside of academia. At the same time, the self-conception of future studies shifted from merely predicting the future to offering recommendatons for actively shaping it.

Today, futures research relies on a range of "imported" methods, particularly from quantitative social research, ethnography, history, cybernetics, economic modeling, and process management, among others.

Transdisciplinary projects are more demanding than their multidisciplinary counterparts because non-academic forms of knowledge and perspectives must be considered on equal footing. This includes "tacit" forms of knowledge and experience that may resist easy explanation or even elude conscious awareness. For this reason, transdisciplinary projects may require non-verbal modes of communication such as visual aids or group activities.

Guidelines

1) *Give due attention to cultures of communication:*
Terminology, concepts, and methods that are unique to a given academic discipline help to facilitate understanding between individuals while also reducing complexity. Similary, individuals from the world of professional practice also have their own unique knowledge cultures and methodological approaches. The accommodation of different styles of thinking and communication is thus an important hallmark of the transdisciplinary project. Indeed, this is necessary to arrive at a shared understanding of important terminology and the project's overarching goals.

2) *Identify and address problems in a participatory process:*
In a transdisciplinary project, project participants may have divergent views of the problem. Part and parcel with this tendency is the inclination for specialists to view their way of conceiving the world as the most valid. Accordingly, at the beginning of a project, participants should explicitly formulate and share their understanding of problem. In a subsequent step, the problem must be defined or "structured" in a manner that is amendable to examination, yet without losing sight of important complexities. If divergent viewpoints are not identified and reconciled early on, then the project will lack a sound basis for collaborative activity, as the participants will fail to operate on the "same wavelength."

3) *Moderate the process:*
Schedules, regular reporting, recurring meetings are the three most important instruments for monitoring and managing the progress of a transdisciplinary project (see "Project and Process Management"). Regular reporting intervals, written updates from project groups, clear project milestones, and recurring meetings (e.g. coupled with reporting dates) all provide traction to keep the project moving forward. The project meetings should permit time for discussion and, if possible, should extend over two days. Meetings and other project events play an essential role for exchanging knowledge and monitoring emerging findings.

Detaching the individuals participating in a project from their respective "home" organizations to enable closely coordinated joint work for the duration of the project will help to reduce the divergence between professional perspectives. Conversely, if the project partners only meet from time to time, individual indentities will tend to remain stronger. Furthermore, the meeting locations are not without significance: shared spaces—even if only used for the purpose of annual meetings—can help to foster a common identity and willingness to collaborate.

4) *Assign clear responsibilities to individual project participants:*
   The clear and unambiguous assignment of tasks and responsibilities to project partners is an important factor for the overall success of a project. One helpful practice is to identify project participants—including their tasks, responsibilities, and tools—in an organizational chart. This chart, which should be mutually defined, helps to underscore that the project is a joint activity that relies on inputs from everyone involved.
5) *Take advantage of new moderation methods and visual communication tools:*
   To ensure that project partners work together as equals, researchers should avail themselves of novel methods of facilitation such as collegial consultation, open-space technology, appreciative inquiry, and futures workshops. Visual forms of communication, including graphics, drawings, and images, are also important for facilitating an understanding of complex problems. One useful method in this regard is *constellation analysis*, which was developed to enhance collaboration in transdisciplinary research projects. Constellation analysis helps individuals to understand their role in a project while also rendering problems and tasks more transparent. It does this by presenting the elements of a project in a visually compelling and intuitively understandable manner.
6) *Engage in effective knowledge managment:*
   To be sure, knowledge can take many forms: in addition to explicit factual knowledge, including statistics or project data, there are many types of tacit knowledge, from practical know-how to experience-based intuition. Organizations—like individuals—have specific knowledge resources and specific ways of using them.
   In a transdisciplinary project, the knowledge assets held by individuals and organizations must be shared, mutually understood, and merged to generate new knowledge. To this end, participants must make their factual knowledge accessible to all partners while rendering their tacit knowledge explicit. In this connection, excessive reliance on shared databases or project intranets can lead to online data repositories that are marginally used and of little benefit. For this reason, it is important to introduce a knowledge management process that effectively gathers and organizes qualitative information in a project. Wherever possible, this management process should simplify interrelationships while prioritizing easy-to-understand images, infographics, and diagrams over written forms of communication. In this way, futures reseachers should consider holding an introductory workshop on project-related knowledge management in order to arrive at a mutually defined process for the project.

7) *Integrate knowledge in an effective manner:*
The quality of knowledge integration in a transdisciplinary project depends on numerous factors, including the joint definition of key concepts; the use of generally understandable language without technical jargon; a reliance on visual and creative forms of expression and depiction; an effective mix of concrete and abstract representations; trusting and open communication; the clarification of expected knowledge contributions; and the management of knowledge assets. These factors should be expressly considered by the project team.

8) *View transdisciplinary cooperation as a learning process:*
In addition to the production of knowledge about the future, *transdisciplinary learning* is a discrete subgoal of the transdisciplinary project. The success of this learning will hinge on the open exchange of ideas, knowledge, and viewpoints between project partners. Accordingly, the goal of mutual learning should be addressed at the beginning of the project and reflected on at regular intervals.

Common Shortcomings and Pitfalls

A. *Failing to define clear terms of reference for participants:*
Attention should be devoted at the beginning of the project to discussing and clarifying how the responsibilities of individual project partners fit into the overall project. The participants may become frustrated or lose motivation if expectations are unclear (who sends what to whom?); if work is duplicated unnecessarily; or if important activities are overlooked.

B. *Failing to jointly define the problem:*
Transdiciplinary projects often have tight deadlines, limiting the amount of time available for participants to exchange ideas. As a result, there is a risk that participants will "go their own way," approaching the project with methods customary to their discipline, but poorly attuned to the overall project. If not corrected early on, this dynamic can lead to poor communication, misunderstandings, and conflict—and may even sabotage the entire project.

C. *Conflating forms of assessment:*
In a transdisciplinary project, various types of assessment are generated and exchanged, including assessments regarding: the source of a problem and the factors that govern it; future states and their attainment; and opportunities for social, technical, or economic change. Conflicts or frustration may arise

if participants fail to clearly identify the type of assessment that is at issue. Problems in this area can be avoided by clearly defining research tasks and by arriving at a common understanding of key concepts at the outset of the project.

D. *Underestimating additional effort:*
Project participants may fail to appreciate in advance that the integration of knowledge in a transdisciplinary project is an extremely time-consuming process. As a result, there may be insufficient time available to explore new ideas, review findings, correct mistakes, or discard intermediate outputs. As a consequence, the knowledge integration process may stagnate, run into roadblocks, or suffer setbacks.

E. *Selecting a poor physical location for cooperation:*
One common pitfall is to overlook the importance of a suitable location for joint project activities. In particular, there are considerable benefits associated with having a fixed site for recurring meetings in which the participants can feel at ease and develop a strong rapport. Constantly changing meeting locations, each with their own deficits (e.g. inadequate seating, offputting decor), can make it difficult to establish a productive working atmosphere or culture of cooperation.

Illustrative Example

On a pleasant June day a research team convenes for a project workshop. Funded by NATO's Science for Peace program, the project is devoted to studying future technological threats to human rights and democracy in Europe. The purpose of the two-day workshop is to present and discuss four scenarios on technological threats that could emerge between 2030 and 2050. During the first phase of the workshop, the four members of the research team—two academics, one industry representative, and one government official—present four different scenarios. This is followed by a collective discussion of each scenario's threat potential, plausibility, and probability. The project brings together representatives from academia, police forces, data protection organizations, and robotics and biotechnology manufacturers. Various representatives from intelligence agencies in NATO countries are also in attendance.

"To make sure we speak freely, the Chatham House rules will apply to this workshop," the group's introductory speaker, a professor of technology law, announces. "If you believe that, I have a bridge to sell you," mutters Frank, a researcher who has just finished a study on how European police forces manage sensitive data. As a result of this research, Frank has a dim view of European police forces, but he is determined to keep his opinions private, as he views the workshop as an excellent networking opportunity.

In the run-up to the event, the intelligence agency participants asked to remain anonymous. Accordingly, instead of proper names, the list of attendees features "intelligence representatives" from various countries, including Poland, Turkey, and Armenia. Despite this emphasis on anonymity, the intelligence officials exchange business cards with numerous individuals on day two.

Another attending researcher, Sarah, is shocked by the matter-of-factness with which an engineer from a British robotics company presents on the use of weaponized drones. "Is it okay to talk like this in public?" she wonders. Sarah notices that her colleague Misha, an Isreali who has previously cooperated with defense manufactuers, seems totally unfazed.

The research project was originally granted funding because it appeared to be a unique opportunity for collaboration between public administrations, companies, data protection specialists, and research institutions. In the funding application, the project partners pledged to share the work equally, and collaborate as equal partners. However, many of the non-academics on the project team are not accustomed to writing things down systematically. While some have past experience collaborating in reseach projects, they prefer to leave the desk work to the "eggheads." Due to time pressure, the project coordinator has authored most of the written material. Accordingly, not all of the scenarios have been developed with the same level of detail prior to the event. Noticing that the quality of the presentations is mixed, Karen, a federal police officer, decides to engage a few of the company representatives in conversation. "Quiet please—the break is over and we're starting again," the coordinator interrupts. "First, please read the two scenarios that were just distributed as handouts."

Over the next three years, the work is in fact distributed equally between the participants, and at the annual project meetings, everyone knows what is expected of them. Karen presented reports on the tools used in her department, such as the Technology Radar method. Over the course of the three-year project, the project participants develop a concrete scenario for how a democratic state can remain resilient to technological threats. Without the cooperation of partners from domains of data protection and civil society, it would have been difficult to discuss operations for counteracting technological threats while also preserving fundamental rights in a free society. Ultimately, the project is successful in furnishing a vision for a self-confident, risk-tolerant, and resilient democratic state.

Further Reading

Bergmann, M., Jahn, T., Knobloch, T., Krohn, W., Pohl, C. & Schramm, E. (Eds.) (2010). Methoden transdisziplinärer Forschung – Ein Überblick mit Anwendungsbeispielen. Campus. [English edition: Bergmann, M. et al. (2021). Transdisciplinary sustainability research in real-world labs: success factors and methods for change. *Sustainability Science*, 16, 541–564.]

Brandt, P., et al. (2013). A review of transdisciplinary research in sustainability science. *Ecological Economics*, 92, 1–15.

Defila, R., Di Giulio, A., & Scheuermann, M. (2006). *Forschungsverbundmanagement: Handbuch für die Gestaltung inter- und transdisziplinärer Projekte Management transdisziplinärer Forschungsprozesse*. Birkhäuser.

Hadorn, G. H., et al. (Eds.). (2008). *Handbook of transdisciplinary research*. Springer.

Mittelstrass, J. (2005). Methodische Transdisziplinarität. *Technikfolgenabschätzung: Theorie und Praxis*, 14, H2, 18–23. [English edition: Mittelstrass, J. (2011). On transdisciplinarity. *Trames*, 15, 329–338.]

Pohl, C., & Hirsch Hadorn, G. (2006). *Gestaltungsprinzipien für die transdisziplinäre Forschung. Ein Beitrag des td-net*. Oekom.

Schmithals, J., Loibl, C., Dienel, H.-L., & von Braun, C.-F. (2011). Kleines Einmaleins inter- und transdisziplinärer Forschungskooperation: Anspruch und Wirklichkeit in der Kooperation zwischen Wissenschaft und Praxis. Empirische Befunde und Handlungsempfehlungen. *Briefe zur Interdisziplinarität*, 8, H2, 3–96.

Schophaus, M., Schön, S., & Dienel, H.-L. (2004). *Transdisziplinäres Kooperationsmanagement: Neue Wege in der Zusammenarbeit zwischen Wissenschaft und Gesellschaft*. Oekom.

Part II
Standards of Group 2: Good Research Practice

Lars Gerhold and Elmar Schüll

The standards presented in this group emerge from the points of difference between futures research and other ways of considering the future.

Futures research diverges from other techniques for generating statements about the future in its adherence to scientific methods, including the transparent documentation of the research. In contrast to the visions of the future offered by science fiction, astrology, prophecy, or bureaucratic planning, futures research is committed to the general principles of good research practice. Compliance with the standards presented in this group lends scientific credibility to the findings of futures research.

Studying the future brings with it special epistemological and scientific demands. These demands render some standards in empirical research inapplicable. Fortunately, numerous important hallmarks of good scientific practice also apply to futures research, including the sensible handling of the research topic (see "Objectives and Framework Conditions") and the use of a transparent, systematic approach that third parties can review and validate (see "Transparency"). Every research method offers a range of possibilities, with its own advantages and disadvantages. The methods applied in a given case ultimately depend on the purposes of research and associated research topic (see "Method Selection"). Method selection is a particularly important concern, as it directly impacts the value of the results.

Closely related to method selection is the theoretical underpinning of a project. This pertains to the project's fundamental epistemological standpoint and the associated theories of the research subject, including assumptions about mediating factors and why they are relevant. While researchers have a certain degree of freedom when selecting their theoretical perspective, this issue is of great import for the research process (see "Theoretical Foundation").

Futures Research that aims to be scientific must be carefully conceived, engage in intelligible argumentation, and use terminology in an unambiguous manner (see "Producing Quality Research"). It is important that researchers describe the scope and methodological limits of their work when communicating their findings, as misinterpretation may otherwise occur. Moreover, they must subject their results to review by the scientific community. Of course, futures research often takes the form of contract research with a strong practical orientation. Yet if futures research is to be scientific, it must contribute to scientific discourse as well (see "Scientific Relevance").

Envisioning the future—including the desirability of potential developments—is a considerable challenge. What makes it more difficult is that most futures researchers explicitly or implicitly inject normative considerations into their work, whether in the form of personal values or societal goals. In order to mitigate the risk of findings being skewed by normative presumptions, it is crucial that researchers critically examine their own values and perspectives.

Futures research is characterized by an ambition closely associated with the Enlightenment: the emancipation of humankind and the spread of freedom. And like other scientific disciplines, futures research must strive to examine ambiguities, answer questions, and fix errors by generating new knowledge. These behaviors belong to the central tenets of good research practice (see "Code of Conduct and Scientific Integrity"). It may seem obvious that apodictic certainty, exaggeration, or dishonesty is detrimental to science—but, alas, it is not. With its orientation to the needs of funders outside science, futures research requires special reminding of the scientific ethos, however tedious this may be at times.

By adhering to the standards presented in this section, futures researchers are better equipped to develop scientifically rigorous images of future that can furnish a sound basis for action.

Objectives and Framework Conditions

Kerstin Cuhls

Summary

Defining the objectives and framework conditions of a futures research project is one of the key tasks in project design, and should be performed at the earliest possible stage. A clear formulation of objectives is essential because it is otherwise impossible to determine whether results have fulfilled intended purposes. Clear objectives are also important given the risk of contradictory aims or potentially hidden agendas. In a similar vein, the background and framework conditions governing the decision to undertake a project should be illuminated, as these conditions impose resource and methodological constraints.

Essentials

At the outset of a research project it is crucial to define objectives, roles, and anticipated results. While the importance of these activities for a transparent research process is generally recognized, researchers may forget to clearly define objectives, or regard them as already obvious. These aspects should be kept in mind during project planning, when managing interfaces (i.e. when results flow into a subsequent stage of the project), and during the subsequent implementation of findings. Each stage of the research process may have its own objective,

---

K. Cuhls (✉)
Competence Center Foresight/Center for Asian and Transcultural Studies (CATS), Japanology, Fraunhofer Institute for Systems and Innovation Research ISI/Ruprecht-Karls University Heidelberg, Karlsruhe, Germany
e-mail: kerstin.cuhls@isi.fraunhofer.de; kerstin.cuhls@zo.uni-heidelberg.de

and objectives are important for the interpretation of the results within a given domain or set of circumstances. In this way, the cascade of interrelationships between overarching objectives and sub-objectives should be elaborated in detail and considered on an ongoing basis.

Objectives may change during the course of a project. When this occurs, it is important to take knock-on effects into account – for example, by making corresponding changes to initial hypotheses or the methodological framework (Diekmann 2001; Flick 2006). The objectives of futures research or foresight processes should always be credible, useful, and comprehensible (see Steinmüller 1997, p. 63). Furthermore, researchers should embrace the fact that objectives may undergo transformation. Indeed, an "ideal" and linear project flow without any iterative loops is rarely found in practice.

The organizers of a research project may have personal objectives that they have not explicitly considered. Research partners, for example, generally have specific perspectives and expectations. Project funders, for their part, may be focused on the economic or reputational benefits they hope to attain. This is particularly true in the private sector. From a scientific perspective, key concerns include advancing knowledge and conducting research in concordance with best practice guidelines.

In order to achieve objectives and to adapt the process to the requirements of the client and relevant framework conditions, normative expectations and perspectives should be considered in a conscious and methodically deliberate manner. While it can be difficult to balance normative expectations with scientific rigor, it helps to make objectives as transparent as possible, in order to allow a clear evaluation of goal achievement. Accordingly, researchers should explicitly formulate the criteria to be used to measure the fulfillment of objectives (see "Transparency"; DeGEval 2008).

In line with the foregoing, when planning a research project, the following guidelines should be considered:

Guidelines

1) *Define objectives and sub-objectives:*
   At the start of the planning process for your research project, clearly define the goals of the entire research process, including all sub-processes, ideally using the cascade mapping technique.

2) *Consider obvious and unconscious expectations:*
   In addition to mutually agreeing on the aims of the process with all participants, strive to render transparent conscious or unconscious but not explicitly stated expectations.
3) *Define purposes:*
   Grunwald describes the purpose-oriented nature of statements about the future, dividing them into "descriptive purposes," "pro-active purposes," and "desirable future developments" (see Acatech 2012, p. 21). In line with this insight, the scientific objectives of research should be adapted to the non-scientific purposes of the project.
4) *Describe the status quo:*
   Be sure to explicitly formulate a mutual understanding of the *status quo* situation in collaboration of the funder and the team carrying out the research. This description may be influenced by the subjective values of participants or the time horizon under consideration, the topic itself, or the available resources.
5) *Strive for clarity when formulating objectives:*
   Objectives and sub-objectives should be defined so that they are clearly understood and interpreted by all participants.
6) *Engage in flexible planning:*
   Project planning should remain flexible enough to accommodate the revision of objectives during the research project. In contrast to experiments in the natural sciences, it is often necessary to react to changing conditions during the course of the project.
7) *Clarify relevant framework conditions:*
   Various conditions may impose limitations on the research process. Accordingly, the following questions should be addressed by the project team, if necessary in writing:
   a) What are the objectives of the process? Are there perhaps unspoken, unconscious, or tacit objectives that can be rendered transparent? Are there perhaps "hidden agendas" behind the scenes? What are the original motives for launching the project?
   b) What is the thematic focus of the project? What topics are on the agenda and what is not to be considered at all? What issues will the process concentrate on? Is there political support and back-up for the undertaking?
   c) What time, financial, and human resources are available?
   d) What is the time horizon considered in the process?
   e) Is the research process participatory in nature? How many people (other than research team members) are expected to participate in the study? Should there be different participants at different stages in the process?

f) What external actors (e.g. experts, laymen) should be involved in the research? What are their occupations and backgrounds? How much can their participation contribute to decision-making? Or will the external actors merely bring together existing information and knowledge?

g) What are the expectations for the results? Is the process itself the goal, and thus the desired "outcome," or is a particular "output" expected (i.e. a tangible result such as a report, image of the future, or recommendation for action)?

Common Shortcomings and Pitfalls

A. *Failing to identify objectives:*
One common pitfall is the failure to explicitly state objectives. This may occur when the participants "forget" to discuss the aims of the research – for example, because they assume the objectives are already obvious.

B. *Failing to define objectives in a clear manner:*
A related pitfall is the ambiguous, imprecise, or cryptic formulation of objectives. Formulations that initially appear unambiguous may be interpreted differently or misunderstood by those involved. One way to avoid this problem is to identify potential conflicts of interest in advance (Sanders 2006, p. 149).

C. *Setting unattainable goals:*
You should avoid setting objectives that are potentially unattainable given the background conditions for research. One consequence of excessively ambitious objectives is the need to request different processes, more time, or additional financial resources mid-project. While foresight processes are often harnessed for PR activities, when they are designed as pure research projects, their process design makes them ill-suited for this purpose.

D. *Overlooking potentially hidden agendas:*
Unspoken aims or intentionally hidden agendas may critically sabotage a research project, as they may cause participants to work at cross-purposes, or create poor alignment between research activities and actual aims. As a result, you should attempt to discern the interests of all parties at the very outset of the process and seek to identify hidden agendas.

E. *Adopting conflicting objectives:*
If not carefully defined, sub-goals may contradict each other or overarching objectives. The interrelationships between objectives should be rendered

as explicit as possible. Furthermore, if any one objective is modified, there is a need to consider possible knock-on effects to other objectives. A failure to adequately consider such interdependencies may harm the quality of the research results, cause changes in the directionalities of the process, or produce unnecessary additional work.

F. *Poorly managing the revision of objectives:*
When changes are made to objectives, researchers must be sure to carefully document and communicate such modifications, as well as adjust their methodological approaches, if necessary. In this connection, it is important to ensure that objectives are not constantly revised during the research process, as this may cause the project to grind to a halt and the team may lose sight of their primary goal.

G. *Inadequate communication or publication of results:*
Various factors may impair the accurate and timely communication of objectives and results – for example, if the client cannot or does not want to understand the methodology, or if the funder does not wish to publish the results because they do not serve undisclosed PR aims or because the result contradicts the (unspoken) expectations of the client. Many clients simply demand "the publication of results" – without clearly disclosing the intended usage context or audience. In such cases, it may be helpful for the researchers to publish their results through a secondary channel (e.g. an academic journal or online as a "working paper"). However, it may take considerable extra time to prepare and release an alternative publication format.

H. *Incorrectly assessing the background conditions for research:*
Some of the political, economic, or technological conditions that led to the decision for a futures research process may not be known to the organizers of the project, but nevertheless have an impact on project implementation later on. In some cases, the root cause may be a hidden agenda or an unclear objective. In a similar vein, the organizers may misjudge the abilities of participants, or the financing of individual methods or materials (e.g. database licenses). In other cases, they underestimate the required resources like time and timing of the different sub-steps that address sub-goals.

Illustrative Example

A non-profit foundation that specializes in funding applied SME research in Germany wants to identify and formulate possible topic areas in the run up to its next call for proposals. Accordingly, it commissions a foresight study on possible areas of consideration for applied R&D. The objective of the project is to identify approximately five broad topic areas that could serve as a basis for soliciting SME funding applications.

As the R&D applications are to fall explicitly under one of these topic areas, the official objective of the foresight project is to develop clear selection criteria and conduct a priority-setting and selection process. The topics need to be interesting and important to SMEs, yet also at the initial research stage, with fundamental research questions still unanswered. As the foundation mainly has experience in mechanical engineering, the range of topic areas is to be broadened considerably (for example to new topics from medicine, biotechnology, etc.).

The contractor – a private research institute – recognizes during the initial briefing that the foundation's unspoken but underlying aim is to increase application rates, which have been declining in recent years. The involved researchers thus propose a partially participatory process, as this promises to attract attention and generate additional applications.

In this way, the practical objective is to identify topics that are important to SMEs and that require cooperation with a research institute for the successful development of a prototype or marketable product. The time horizon is 5 to 10 years until first application in a product. For the call, the topic areas are circumscribed in order to enable tangible outputs on a fixed funding budget.

The futures research methods are chosen according to these specifications. While the initial idea is to launch the project with a workshop, due to lack of interest this idea is abandoned in favor of an open process for gathering topics from SME employees in a survey. To this end, a simple survey is designed and distributed by letter, fax, and e-mail. Potential survey participants are selected from companies known to the foundation as well as from public databases.

The collected survey data are analyzed and clustered into groups using the software application VantagePoint in a modified form for pre-clustering. The topics proposed are then re-formulated, re-evaluated, and re-assessed through the administration of a second survey. In the second survey, the participants are asked to assess the importance of each topic, its time horizon, and the measures necessary to create a mature product. Comments of any kind are possible.

Based on the quantitative and qualitative responses that are received, the foundation selects a range of potential topics for closer examination. The research institute then conducts interviews with experts from the domains of business, academia, and politics in order to assess topic viability.

The topics identified in this manner are described in a standardized written format and submitted to the foundation. An explanation of why the topic is considered interesting is attached. From this list of prioritized topics, the responsible advisory board of the foundation selects the topics that will be included in its call for proposals. In this way, the goal of using a foresight process to identify a sufficient number of good-quality topics is achieved. Furthermore, the project addresses the indirect and unspoken objective of increasing application rates. The application rates undergo a significant jump, allowing the foundation to achieve its project funding targets.

References

Diekmann, A. (2001). *Empirische Sozialforschung: Grundlagen, Methoden, Anwendungen*, (7th revised edition). Rowohlt Taschenbuch Verlag GmbH, Reinbek bei Hamburg.

Flick, U. (Ed.). (2006). *Qualitative Evaluation Research: Konzepte*. Umsetzungen, Rowohlts Enzyklopädie im Rowohlt Taschenbuch Verlag, Reinbek bei Hamburg.

Sanders, J. R. (Ed.). (2006). *Handbook of Evaluation Standards: The Standards of the Joint Committee on Standards for Educational Evaluation*, (3rd and updated edition). Verlag für Sozialwissenschaften, Wiesbaden.

Steinmüller, K. (1997). *Beiträge zu Grundfragen der Zukunftsforschung*, WerkstattBericht 21 des Sekretariats für Zukunftsforschung, Gelsenkirchen.

Further Reading

Acatech (Ed.). (2012). *Technikzukünfte: Vorausdenken – Erstellen – Bewerten*. Acatech IMPULS, Springer Verlag, Berlin.

Cuhls, K. (2008). *Methoden der Technikvorausschau – eine internationale Übersicht*. IRB Publishers.

DeGEval—Gesellschaft für Evaluation e. V. (Ed.). (2008). Standards für die Evaluation (4th unchanged edition). https://www.degeval.org/fileadmin/user_upload/Sonstiges/STANDARDS_2008-12.pdf. Accessed 21 June 2021.

Transparency

Elmar Schüll and Lars Gerhold

Summary

Researchers in the sciences have an obligation to transparently document and disclose their assumptions, methods, and results in a manner that allows scrutiny by third parties. Such transparency is essential not only for the acceptance of a project's findings, but also for the general advancement of scientific knowledge.

Essentials

Transparency is a core hallmark of good scientific work. By clearly disclosing all aspects of a project, researchers make their activities and the results of their work amenable to review and critique. Accordingly, futures researchers have an obligation to clearly and honestly describe the problem setting, explain their research approach, disclose their methods of data gathering and analysis, and transparently document their results and conclusions. Transparency enables the scientific community to evaluate the strengths and weaknesses of a research approach and the specific decisions taken in a research project. In this way, transparency lays the foundation for research results to be understood, critically evaluated, and subsequently accepted or rejected.

E. Schüll (✉)
Salzburg University of Applied Sciences, Salzburg, Austria
e-mail: elmar.schuell@fh-salzburg.ac.at

L. Gerhold
Freie Universität Berlin, Berlin, Germany
e-mail: lars.gerhold@fu-berlin.de

© The Author(s), under exclusive license to Springer Fachmedien Wiesbaden GmbH, part of Springer Nature 2022
L. Gerhold et al. (eds.), *Standards of Futures Research*, Zukunft und Forschung, https://doi.org/10.1007/978-3-658-35806-8_8

Guidelines

From the very beginning of a project, researchers should commit themselves to thoroughly documenting their activities. Furthermore, this documentation should be stored in a suitable format and made available to relevant stakeholders. Particularly when a project is of public interest, researchers should strive to publish their findings in professional journals or present them at conferences. Of course, private or political interests may stand in the way of full disclosure in individual cases. However, the scientific obligation to work transparently also applies to private research contracts. In this way, when essential aspects of a research project cannot be made transparent, this must be viewed as a deficiency from a scientific perspective, even when this lack of transparency is unavoidable due to non-disclosure obligations.

The following guidelines should be followed when documenting a project, including individual research steps:

1. *Describe the purpose of the research:*
   One should transparently describe the initial problem situation, relevant questions and objectives, and the desired outputs of the research process. What is the *purpose* of the study – in other words, what does one hope to achieve?
2. *Make assumptions explicit:*
   The assumptions and expectations of the research team should be recorded before the start of the project and compared ex post with the obtained results. Such a comparison can be important for identifying the common problem of prior expectations skewing the research process. A strong commitment to the documentation of assumptions and expectations can significantly enhance the objectivity of a project.
3. *Explain the theoretical framework:*
   Despite the strong practical dimension of futures research, adequate consideration of relevant theory is an important aspect of research work. In this respect, researchers should explain which theories are being drawn upon and why they are relevant to the subject matter of investigation (see "Theoretical Foundations").
4. *Define key concepts:*
   In order to ensure that research questions can be answered in a targeted manner, it is crucial to clearly define concepts and terminology. How should a given term be understood? What other definitions are conceivable? Which concepts have been chosen, and for which reasons? Addressing such issues

is important not only for effective collaboration within the research team but also for ensuring the findings of the study are clearly understood.

5. *Handle data transparently:*
   Explain how data were collected, including relevant methodological steps. If existing data from third parties are used, one should comprehensively account for the sources of these data while also noting any associated limitations.

6. *Explain your results:*
   Shed light on the process by which collected data and observations are interpreted. Which theories and logical operations led to the conclusions? In which ways are the project's findings potentially flawed or open to objections?

7. *Thoroughly document the process:*
   Be sure to comprehensively document all important steps in the project. In the absence of confidentiality concerns, third parties should have a right to inspect project documentation and data. Similarly, researchers should embrace discussion and debate concerning their data, methods, and results.

8. *Stick to essentials:*
   Avoid getting bogged down in excessive detail. In this regard, common sense is key. The fine line between necessary documentation and extraneous nuance should be determined collectively by the research team on a case-by-case basis.

Common Shortcomings and Pitfalls

A. *Inadequate time or resources:*
   Perhaps the most common pitfall is a failure to document the project with sufficient care or detail as a result of time or budget constraints or because of the departure of a staff member prior to a project's conclusion.

B. *Inaccurate or missing documentation:*
   Researchers may fail to provide documentation or discuss important components of the project in an excessively vague manner. Failure to ensure transparency must be considered a deficiency from a scientific perspective, even if necessary due to confidentiality requirements.

C. *Lack of acceptance for criticism:*
   At the end of a study, researchers should reflect on the research process, rather than merely drafting a final report and moving on to the next project. In a similar vein, researchers may view working meetings or conferences solely

as an opportunity to enhance their reputation, and show little willingness to subject the study to critical appraisal.
D. *Improper documentation formats:*
This pitfall involves a failure to properly store and archive data and other project records.
E. *Hyper-transparency:*
This pitfall occurs when the documentation is compiled with the greatest possible depth of detail, such that the essential matters are submerged in a flood of material. This may make it difficult or even impossible for outsiders to access or understand important aspects of the project.
F. *Lack of clarity concerning normative assertions:*
This pitfall involves failing to make a clear distinction between normative positions and objective descriptions. Researchers should refrain from surreptitiously or unconsciously intermixing these two perspectives (see "Modality").

Illustrative Example

A research team obtains funding to study future developments in the domain of infrastructure security. The researchers draw on relevant source literature pertaining to risk management and security policy, as well as studies concerned with anticipated technological change up to 2040. One preliminary assumption of the researchers is that various potential hazards are conceivable, but that no expectable technological developments can offer adequate safeguards. The researchers document their literature review, and evaluate the information thus gathered in terms of its relevance.

After elaborating a methodological design, the research team holds a kick-off workshop to discuss the research questions, theoretical backdrop, and key concepts. Various approaches are considered, and the team ultimately selects a Delphi survey as the most suitable. The discussions during this workshop are recorded, paraphrased, and validated by the research team. During subsequent review of the workshop protocols, various ambiguities are identified. The team realizes that key terms such as "security" and "critical infrastructure" have not been adequately defined and distinguished from other concepts. As a result, follow-up work is performed to address this deficit.

The next step is to develop a survey instrument. During this process, the researchers document which questions and aspects are being included, and for which reasons. They also explicitly note issues left unaddressed, e.g. due to financial or time constraints.

The researchers describe the panel of experts being consulted in terms of their professional backgrounds and demographic characteristics, in order to enable subsequent verification of the panel participants. The Delphi survey is administered using an online survey tool. The gathered data are transparently disclosed to the survey individuals and in the project documentation. Generally, the gathered data are analyzed at a descriptive level. However, the research team also tries to explain the differences between survey waves and panel subgroups using a multivariate analysis approach.

A workshop is then held to assess, interpret, and validate the survey results. The summary of findings generated as part of this workshop contains a plausibility-weighted catalog of possible future developments and risks in relation to critical infrastructure up to 2040. These results are used to formulate recommendations for action for relevant stakeholders – that is, decision-makers responsible for disaster response measures. The final report comprehensibly documents the results and key findings of the project. The individual steps used to generate the recommendations for action are also clearly explained.

The practical findings of the project are presented as a report to stakeholders and in a public lecture. The researchers also publish an article in an academic journal to disseminate the methodological insights that emerged from the project.

Despite an urgent need to devote attention to a new project, the researchers take time to review their documentation and throw out unnecessary records.

Further Reading

Denzin, N. K. (2009). *The research act. A theoretical introduction to sociological methods.* Routledge.
Flick, U. (2019). *An Introduction to Qualitative Research.* 6th Edition. Sage.

Theoretical Foundations 9

Elmar Schüll

Summary

How we understand and explain the world largely depends on our experiences and the expectations and presuppositions that attend them. In science, this phenomenon is known as "theory-dependent observation." Futures researchers can produce high-quality work only if they keep in mind the importance of theoretical models for science and research.

Essentials

An essential characteristic of good scientific research is a sound theoretical foundation. The interpretation of data, the methodology, the choice of the research question, and other research activities should not be based on "common sense," everyday intuitions, or subjective feelings, but on empirical findings and scientific models. "Theory" and "theoretical foundations" are ambiguous, however, because they can describe different aspects of scientific practice. Two are worth noting here in particular:

Theories of science: Over time, various ideas have emerged about what goals and types of knowledge scientific research should pursue. Every theory of science describes a specific way of evaluating data, designing research, choosing methods

E. Schüll (✉)
Salzburg University of Applied Sciences, Salzburg, Austria
e-mail: elmar.schuell@fh-salzburg.ac.at

© The Author(s), under exclusive license to Springer Fachmedien Wiesbaden GmbH, part of Springer Nature 2022
L. Gerhold et al. (eds.), *Standards of Futures Research*, Zukunft und Forschung,
https://doi.org/10.1007/978-3-658-35806-8_9

(see "Method Selection"), and determining the validity of the results. Some theories of science are positivism, critical rationalism, critical theory, pragmatism, and constructivism.[1]

Theories of the research subject: Each scientific discipline has produced more or less generalizable findings regarding its area of research. Theories of the research subject – that is, regarding the subject matter of investigation – vary widely in terms of their generality or validation, and they encompass rules of scientific practice, models, analytical frameworks, meta-concepts, and heuristic approaches. All seek to circumscribe a discipline's area of work meaningfully and accurately. For research that claims to be scientific, theories of the research subject are the orienting frame of reference.

Theories of science and theories of the research subject are analytically distinct, but mutually reinforcing in practice. The former suggest a certain subset of the latter, and both in turn entail a certain repertoire of methods (see Voros 2007, p. 84 f. as well as "Method Selection"). For example, experimental designs have a place within the tradition of positivism or critical rationalism, but less so within critical theory. It is important to reflect critically on these loose relationships in order to avoid regarding a certain approach as the only possible one.

Theories of science and theories of the research subject undergo change over time as they evolve or become modified, supplemented, or replaced by competing theories – typically in the name of progress (e.g. Chalmers 2013 or Kuhn 1989). Theories, therefore, are not timeless, unchanging truths; they are concepts that human beings make and that human beings can change.[2] Futures researchers are well advised to keep abreast of theoretical discussions in the fields relevant to their work.

When deciding on a theory of science and a theory of the research subject, usually more than one reasonable option exists. A choice must be made, although the right choice is rarely obvious and rarely follows from the nature of the investigated research subject. Theories of science encompass different paradigms – elementary views about the meaning and purpose of scientific work. Usually, paradigms are acquired and internalized in many years of academic

---

[1] Chalmers (2013) provides an overview of the history and development of various theories of science. Voros (2007) shows their importance for futures studies. On the uses and limitations of critical rationalism in futures research, see Schüll & Berner (2012).

[2] Theories claim to be generalizable, i.e. to describe mechanisms and relationships that are valid regardless of the time or place. Thus, a theory's degree of generalizability is a criterion for its quality – the more general, the better. But this does not change the fact that theories are devised by human beings, and notwithstanding claims to "timeless" validity, will sooner or later be questioned, supplemented, or revised.

9 Theoretical Foundations

socialization, and are not easily interchangeable.[3] In a sense, theories of science are like articles of faith: they are assumptions about what is correct, true, meaningful, or relevant.[4] The danger comes when researchers regard one theory of science as self-evident, and fail to recognize the utility of other approaches.

When it comes to specific research projects, researchers usually have several theoretical approaches to choose from. For example, those studying the future of higher education can analyze universities as organizations, as performers of specific functions (research, teaching, knowledge transfer, etc.), as social systems, as producers of equality or inequality, or as historical exceptions due to their centuries-long existence (Hechler and Pasternack 2012, p. 9). Whether researchers pursue a theory of organization, a theory of education, a theory of systems, or a theory of function depends on the focus of their work. But even when the focus is clear, researchers still have several approaches to choose from – some complementary, some competing (Hechler and Pasternack 2012, p. 9–39).

Though it is true that the choice of theoretical foundations is rarely self-evident, this does not mean that it is arbitrary. Scientific theories fulfill various functions, and the quality of research depends to a large extent on how they perform. Thus, even when several theoretical approaches are possible, not all of them will be equally useful in the context of a specific project's research subject, knowledge interest, use function, available resources, funders, and audience.

Theories of the research subject help investigators identify important aspects of their work, formulate precise research questions, and develop an appropriate research design for developing hypotheses and analyzing developments (see "Method Selection").

The central function of theories of science and theories of the research subject is to ensure transparency with regard to the research process and the research results. One characteristic that distinguishes scientific knowledge from common knowledge or personal experience is that it has been validated (see Grunwald, 2013 and "Validation by Argumentation"). For the assessment of futures research,

---

[3] "A paradigm may be viewed as a set of basic beliefs […] that deals with ultimates or first principles. It represents a worldview that defines, for its holder, the nature of the 'world,' the individual's place in it, and the range of possible relationships to that world and its parts. […] The beliefs are basic in the sense that they must be accepted simply on faith (however well argued); there is no way to establish their ultimate truthfulness. If they were, the philosophical debates […] would have been resolved millennia ago" (Guba and Lincoln 1994, p. 107; quoted in Voros 2007, p. 75).

[4] Not surprisingly, theories of science have often been hotly contested. The positivism dispute and the value judgment controversy are two prominent examples (see Adorno 1993 and Dahms 2007).

it is crucial that the theoretical foundations be clearly documented. The theoretical foundations of the results permits external criticism and improvement – the conditions of the possibility of scientific progress (see "Transparency").

Guidelines

1. *Use scientific theories to guide research:* It is important that researchers let scientific theories guide the entirety of their work. Theories aid in gaining knowledge, provide orientation, and enable targeted research. Moreover, they help identify the non-scientific interests, expectations, and heuristics that researchers inevitably bring with them.
2. *Stay abreast of the current state of research:* Employing scientific theories productively is no easy undertaking. Of course, introductory literature exists covering the most important theories for each discipline, and researchers can always consult with other colleagues from the scientific community. But theories are not eternally valid truths; they continue to evolve over time and can be supplemented or replaced by other theories. Hence, futures researchers must keep up with the scientific literature and take note of new research relevant to their work.
3. *Do not mistake being scientific for being superior:* The use of scientific theories distinguishes scientific knowledge from other forms of knowledge. This does not mean that scientists know things "better," but rather that they know things "differently" (see Hechler and Pasternack 2012, p. 7). Researchers must keep this in mind, especially in futures studies, where the transfer of scientific knowledge to applied areas is common. The use of scientific theories is crucial for ensuring the quality of research. However, "being scientific" is not a reason to feel superior over project partners or the research audience.
4. *Reflect critically on the scientific theories you have chosen to use:* Researchers must reflect critically and explicitly on their own scientific theories. They must make transparent the theoretical foundations of their results and the type of validity they claim.
5. *Ensure theoretical diversity and scientific rigor:* Researchers should consider the diversity of theoretical traditions and the relativity of their own position. This does not mean that "anything goes," however. Scientific rigor is crucial, regardless of the scientific paradigm. Of course, each theoretical approach typically brings with it a mandatory framework for subsequent research decisions. But familiarity with the variety of scientific theories prevents researchers

from overvaluing their own position and undervaluing other, equally possible approaches.
6. *Document initial assumptions:* One way to monitor one's own work is to write down ad hoc expectations at the outset of the research. At the end of the study, the notes can be checked against the outcome to see whether the research produced new findings or whether it merely served to legitimize pre-existing expectations. Of course, the research may corroborate initial assumptions that were already correct. (This is especially likely in the case of experienced researchers.) But it is also possible that the initial assumptions bent the entire research process toward the desired results. This is why a critical examination of the initial assumptions is crucial.

Common Shortcomings and Pitfalls

A. *Forgoing science and theory:* The fact that futures research is usually applied contract work sometimes leads investigators to forgo scientific knowledge and theory. But in so doing, they fail to recognize that perceptions, decisions, and interpretations are fundamentally theory-driven. Futures studies without a theoretical framework – whether due to the wrong priorities or the lack of time and expertise – allow everyday theories, ad-hoc assumptions, or generalized individual values to gain the upper hand. "Flying blind" in this way typically results in a simplistic approach with little gain in knowledge. The problem is not only theoretical: futures research can be highly relevant for decision-making and action (see "Aligning Research with Ambitions for Action"), and if recommendations are based on poor research, real problems can result (see, for example, Rust 2008, p. 50 f.).
B. *Failing to understand the added value of scientific theory:* Researchers may mistakenly believe that scientific theories are merely time-consuming, complicated, and burdensome. In truth, theoretical knowledge is what distinguishes futures researchers from their (typically well-informed) audience. The deft use of scientific theories makes for more rigorous and focused work, enabling high-quality research even when time and resources are scarce.
C. *Failing to reflect critically on a theory of science:* Researchers may fail to base their work on an explicit theory of science, resulting in an unclear research question or vacillating goals. Especially in futures studies, researchers must be aware of the scientific validity of the results (scenarios, forecasts, present analyses, trend extrapolations, etc.).

D. *Using methodology to compensate for gaps in theory:* In future studies, researchers tend to neglect theory and compensate for the resulting uncertainty with extensive methodological effort. But no matter the effort or the quality of the methodology, results do not speak for themselves; they require a scientifically sound interpretation.

E. *Succumbing to scientific superficiality:* Because futures researchers are often tasked with solving practical problems, the usual approach in science – investigating a narrow research question informed by a single theory of the research subject – rarely suffices. Futures researchers usually require an interdisciplinary approach (see "Interdisciplinarity"). Interdisciplinarity represents a special challenge and, if properly implemented, a special achievement. The key lies in understanding the discussions of scientific theory in all disciplines relevant to the research. Otherwise, researchers may succumb to scientific superficiality – saying something about everything without really knowing anything.

Illustrative Example

With the transitional regulations restricting the free movement of workers for states that joined the EU in 2004 and 2007 about to expire, the Austrian government and the City of Vienna want to obtain expert opinion on the future flow of migrants. The aim is to collect estimations of the inflow and outflow figures over the next 10 years so that officials can respond to the social and economic consequences in a timely manner.

The government publishes a tender for the study and gives the contract to a small private-sector research center based in Vienna, which has already carried out multiple projects in cooperation with the University of Vienna and other institutions of higher learning.

The study is divided into two parts: the first focuses on estimating future migration flows; the second, on recommendations for integrating new immigrants. Immigration is a sensitive political issue because the general population can perceive newcomers as a threat rather than a boon. Many assume that immigrants will exploit the welfare state and force local workers from the labor market. Almost all the government's models are designed accordingly: only after immigrants prove that they are "useful" members of society can they can hope for legal and social recognition.

9 Theoretical Foundations

The demographers in the research team consider several forecasting models to estimate the magnitude of future migration. It soon becomes obvious that the idea of demography as a stronghold of accurate and reliable forecasts is wrong. Migration, in particular, is the strongest element of uncertainty in demographic forecasts and has the greatest short-term effect on the destination country's population and social system. The uncertainty is also emphasized in communications with the authorities of the City of Vienna and the country of Austria. The researchers note that, according to recent studies on the accuracy of demographic forecasts, a prediction using current models will be inaccurate. The research team proposes to address the forecast uncertainty by identifying the main factors influencing migration behavior and designing several migration scenarios. They will then draft suitable recommendations for each of the scenarios, covering a broad spectrum of possible outcomes.

A major challenge facing the researchers is that migration is influenced by multiple factors – at once political, economic, and societal. So they assemble a team covering the spectrum of disciplines required for their work. To ensure that their approach is in line with the latest research, the investigators review the most recent literature in each discipline. They pay particular attention to the theoretical assumptions in political science, social science, and economics.

Drawing on various theories, the research team develops a complex model that can approximately estimate past migration flows. They publish the model in a peer-reviewed journal. There it becomes clear that, as with all models, theirs cannot capture the entire range of factors determining the research subject. The data required for the scenarios are either publicly available or provided by state authorities. Accordingly, the researchers' methodological effort is limited to the evaluation of the data.

The theoretical and normative framework for the study's recommendations is based on Axel Honneth's (1994) theory of recognition and a number of spin-off concepts that have proven successful in other countries. Drawing on that work, the researchers suggest that immigrants receive speedier legal recognition and that spaces and institutions be introduced in which they can express their needs. They expect that the approach will increase societal cohesion and promote social and economic inclusion.

Government officials are initially skeptical: the recommendations are at odds with prevailing public and political opinion. But considering the study's model and the best-practice examples, they come to see the conclusions as logical and appropriate. In the end, they are pleased with the work and value the new perspectives on migration and inclusion. But they remain uncertain whether implementing the recommendations will be politically feasible.

References

Chalmers, A. F. (2013). *What is this thing called science?* 4th ed., Maidenhead: McGraw-Hill.
Dahms, H. J. (2007). *Positivismusstreit: Die Auseinandersetzung der Frankfurter Schule mit dem logischen Positivismus, dem amerikanischen Pragmatismus und dem kritischen Rationalismus.* Frankfurt/Main: Suhrkamp.
Grunwald, A. (2013). Wissenschaftliche Validität als Qualitätsmerkmal der Zukunftsforschung. *Zeitschrift für Zukunftsforschung*, 2. http://www.zeitschrift-zukunftsforschung.de/ausgaben/jahrgang-2013/ausgabe-2/3694. Accessed: 28 May 2021.
Honneth, A. (1994). *Kampf um Anerkennung: Zur moralischen Grammatik sozialer Konflikte.* Frankfurt/Main: Suhrkamp.
Rust, H. (2008). *Zukunftsillusionen: Kritik der Trendforschung.* Wiesbaden: VS-Verlag für Sozialwissenschaften.
Schüll, E. & Berner, H. (2012). Zukunftsforschung, kritischer Rationalismus und das Hempel-Oppenheim-Schema. In R. Popp, Reinhold (Ed.), *Zukunft und Wissenschaft: Wege und Irrwege der Zukunftsforschung.* Berlin: Springer-Verlag.
Voros, J. (2007). On the philosophical foundations of futures research. In P. Duin (Ed.), *Knowing tomorrow? How science deals with the future*, 69–90. Eburon Academic Publishers.

Further Reading

Adorno, T. W. (1976). *Der Positivismusstreit in der deutschen Soziologie: Zu Werturteilsdiskussion und Positivismusstreit.* Darmstadt: Luchterhand.

Method Selection

Lars Gerhold

Summary

The empirical methods selected by researchers should be suited to the project's purpose, subject matter, and conditions (including time and resources). Various methodological approaches can be combined if this facilitates a deeper understanding of the issue being examined or otherwise helps to answer the questions at the heart of the project. If different methods are used, the relevance of each method to the project's purpose must be explained. One advantage of using several methods is the ability to compensate for specific weaknesses or obtain more comprehensive results. Yet to this end, researchers must have a sufficiently large pool of methods to choose from. Furthermore, the methods assembled must be justifiable in a transparent fashion. This is an important prerequisite for audiences in the domains of academia, business, and government to understand and accept the findings of the project, including associated recommendations for action.

Essentials

The range of research methods and techniques used in futures research is extremely broad (see Glenn & Gordon, 2009, among others). In part, researchers rely on established methods of empirical social science, including quantitative surveys and qualitative interviews, adapting them for the needs of futures studies

---

L. Gerhold (✉)
Freie Universität Berlin, Berlin, Deutschland
e-mail: lars.gerhold@fu-berlin.de

© The Author(s), under exclusive license to Springer Fachmedien Wiesbaden GmbH, part of Springer Nature 2022
L. Gerhold et al. (eds.), *Standards of Futures Research*, Zukunft und Forschung, https://doi.org/10.1007/978-3-658-35806-8_10

(see Steinmüller, 1997, p. 28). In addition, they rely on a wide variety of methods to solicit knowledge from practicing professionals and academics in other disciplines, including techniques to moderate discussions or generate ideas. Last but not least, future researchers may develop new methods or adapt them to the requirements of the futures research project.

All too frequently, however, futures researchers fail to consider or address method selection in a deliberate and careful manner. This shortcoming may tarnish the research design process. Alternatively, it may impact the later communication of the project's findings, if data collection and analysis are not discussed transparently. Indeed, researchers often neglect to explicitly consider or discuss whether the methods used are appropriate for answering the questions that motivated the project. The goal of future studies is to elaborate statements about the future that are consistent with the empirical knowledge base as well as standards of practice in the discipline (see "Images of the Future"). This underscores the importance of selecting a suitable set of research methods.

While the methodological diversity of futures research can be viewed as an asset, a closely associated pitfall is the deployment of methods for seemingly arbitrary reasons. Indeed, a vague conviction that "more is better" is not sufficient to justify a motley assortment of research methods. Instead, the method selection should emerge coherently from the research question and aims of data collection and analysis. The goal should be to arrive at a rationally justified selection of methods.

Ideally, research methods are selected based on extent to which they are useful to adequately address a research question. While no researcher can know or use all available methods, procedures, and techniques, it is essential to familiarize oneself with the specific strengths of as many futures research methods as possible. To this end, researchers would be advised to engage with the relevant source literature, in order to obtain a grasp of the available methods (e.g. Glenn & Gordon, 2009; Steinmüller, 1997), as well as to follow discussions regarding methods in scientific journals.

When learning about new methods, one should explicitly consider its disciplinary origins. Many methods used in futures research originally come from other established fields, such as sociology, psychology, or political science. Accordingly, one should seek to appreciate the existing body of discourse surrounding a given method, including discussions about its strengths and weaknesses.[1]

When combining methods within a research process, there is also a need to consider the relationships between methods. If selected methods have specific requirements or produce specific results, their combination can generate supplementary benefits, but this is not always the case. The goal of using various methods in a project with a mixed method design should be to facilitate a deeper and more comprehensive understanding of the topic under consideration. In particular, it may be possible to offset the weaknesses of individual methods with the strengths of others (Gerhold, 2012).

Furthermore, in a mixed method design, one must consider that every methodological step is related to the others. Depending on the design of the study, one can select a sequential design or a parallel design. In a *sequential design*, different methods are used one after the other. For example, an explorative brainstorming method may be followed by a quantitative Delphi study. The separate methodological steps each lay the basis for the following step. The advantages of a sequential design are that findings from a first methodological step can be used for the development of further steps. However, this makes them significantly more time-consuming. Furthermore, the data may be subject to larger fluctuations (due to the evolving perceptions of survey participants, or the influence of significant political events, to name but two examples). In a *parallel design*, by contrast, one examines an issue from different empirical and theoretical perspectives simultaneously, without sequential interdependencies.

When developing images of the future on the basis of hypotheses, projections, and supposition, it is necessary to identify potentially relevant interconnections between the sets of data being considered. The generation of an overall interpretation or synopsis can serve data validation ("the different data point to the same result or to different results"). Alternatively, such a synopsis can highlight the complementarity of the data sets ("the data complement each other coherently while reflecting different research perspectives"). The goal of validation can only be pursued if the methods are comparable. By contrast, divergent methods (such as scenario techniques and the Real-Time Delphi method) can only lead to complementary results.

Overall, responsibility for the quality of a project's findings resides with the research team, and when something goes awry, one cannot simply blame the

---

[1] An example of this is the established discourse on qualitative methods in the social sciences, as documented in the Forum Qualitative Social Research (http://www.qualitative-research.net).

methods employed, no matter how complex the study's design. By the same token, the data set produced by a given research method does not stand on its own, but must be analyzed and interpreted in light of the underlying research question and its theoretical framing (see "Theoretical Foundations").

Guidelines

1. *Method Selection:*
   Methods should be chosen in alignment with the purpose of research and issue being studied. One should refrain from selecting methods based on personal bias, preference, or one's own strengths. The method selection should be justifiable, comprehensive, and consistent with a view to the purpose of research. One should explicitly explain why one or more methods are being used, and what the benefit of each method is for answering the question at issue (see "Transparency"). If several methods are combined, one should explain the associated intentions, and how the methods will be integrated.
2. *Literature:*
   The methods of futures research have specific strengths in distinct areas. One should reflect on these strengths when planning the project and when collecting and analyzing data. To this end, one should survey the current literature on relevant approaches, theories, and methods. Be sure to explicitly reflect on and discuss the use of methods in light of this literature.
3. *Theoretical Understanding:*
   Define the research team's theoretical understanding of the matter under investigation or at least the thematic framework of the study in order to make method selection comprehensible (see "Theoretical Foundations"). Explain how methods are informed by a given set of theoretical relationships (or perhaps even posit these relationships).
4. *Mixed Method Design:*
   When combining methods, you should determine the objective prior to undertaking research. This is necessary to assess in retrospect at the stages at which research objectives were achieved (or could not be achieved) based on the methods chosen. In the final project report, you should explicitly explain which results have been included, based on which methods. Furthermore, you should transparently account for why certain results enjoy a preferential status, if applicable.

5. *Methodological Structure:*
   Outline how single methodological steps build on one another, and if they are based on what came before. This is necessary, for example, when qualitative data are transferred to questionnaires and rating scales. Any potential loss of information must be documented and its acceptance justified. For example, you should explicitly state that only selected facets of a problem are being addressed due to questionnaire length limits.
6. *Adaptation of the Research Design:*
   During the research process, it may be necessary to change or adapt the composition of the methods used. This may become necessary due to a change in goals or a modification of the research question. Changes to the methods used or combined should be clearly disclosed and explained.
7. *Reflection:*
   The final report should explicitly reflect on the methodological design, including its strengths and weaknesses. It must become clear whether the chosen methodological steps were able to answer the research question and where the findings have limitations (see "Scientific Relevance").
8. *Disclose Problems:*
   Disclose practical problems and errors in the use of the research methods. This is the only way to learn from experience for the benefit of subsequent and similar research settings. Common problems include excessively small subject pools, technical errors (e.g. unavailability of an online survey) or difficulties in identifying and recruiting experts.

Common shortcomings and pitfalls

a) *Choosing methods based on personal preference:*
   A method should not be selected simply because of prior familiarity, client requirements, or a strategic effort to obtain funding. Similarly, methods should not be chosen due to the composition of the research team while disregarding their suitability for the project. Part and parcel with this pitfall is a failure to take into consideration a sufficiently broad range of possible research methods.
b) *Non-compliance with methodological standards:*
   Yet another common error is to misunderstand methods or misapply them due to lack of consideration for the relevant academic literature. In this connection, one may present or combine methodological steps incorrectly.

c) *Using methods just because others do:*
Methods should not be tossed into the mix because they are "all the rage" and widely used, without considering their specific relevance for the purpose of research. Rather, methods should be chosen based on their intrinsic attributes and capabilities.

d) *Change in research design:*
Due to a lack of adequate planning, it may become clear during the course of a study that the selected methods are not suitable for achieving the research objective. Yet one should not simply add new methodological steps to compensate for planning errors. Rather, the expansion of the method toolbox should directly serve quality enhancement. Furthermore, one should explicitly consider in advance the implications of retaining the original methods, before changing course mid-stream.

e) *Lack of supplemental benefit:*
One should not blindly presume that adding additional methods will generate better or more valid results. Rather, the supplemental benefits gained should be explicitly discussed and established, and care should be taken to ensure the methods complement each other.

f) *Misinterpretation:*
This pitfall occurs when researchers directly compare data sets of different data quality or use them for mutual validation, even when they are unsuitable for this purpose (e.g. "the Delphi study did not corroborate the results of the cross-impact analysis").

g) *Missing key figures:*
This shortcoming involves a failure to mention important parameters of the collected data (e.g. sample size, sampling method, expert selection technique, sources included, statistical properties, or selected time horizons, among others).

Illustrative example

A university research group is preparing a research project on the topic of airport security in 2030. The aim of the research project is to develop scenarios that depict both probable as well as socially and politically advantageous developments in this domain. The research object is being addressed by an interdisciplinary group of academics and professionals from the fields of engineering, the social sciences, law, and airport security. The overall aim of the research project is to develop a security strategy for the next 10 years that considers technical innovations while also reflecting on the ethical and legal dimensions of their use.

The theoretical framework is informed, on the one hand, by a strand in the literature on securitization and, on the other hand, by the normative aim of mitigating dangers to society by deploying advanced security technologies.

In order to gain a deeper understanding of the subject matter under consideration, the research team concludes that a combination of different methods is required. They ultimately decide the project will combine (I) explorative expert interviews; (II) a quantitative Delphi survey; and (III) a scenario development process. Data collection is to be carried out in several sequential phases, i.e. one methodological step will form the basis for the next.

(I) In order to map the research project in its thematic and theoretical breadth, the main aspects of the interdisciplinary approach are to be developed using explorative expert interviews. Qualitative interviews are conducted with engineers to describe the technical options for airport surveillance; with ethicists to explain the moral and value-based view of surveillance technologies; and with practitioners describing their daily routines and experiences of working at security gates, analyzing video data, and screening passengers. The goal of the interviews is to expand the conceptual space of airport security and to obtain indications of which facets of the problem should be explored in more detail.

(II) Based on the expert interviews, substantive areas of emphasis are developed for the Delphi survey. The topics of investigation are: Acceptance of technical measures by airport users; implementation of invisible sensor technologies and hazardous substance detectors in the various security zones of the airport; changes in social values with regard to security in public spaces; changes in EU legal frameworks; further development of IT-based monitoring procedures (video tracking, face recognition). Based on a comprehensive literature research and the interview data, questionnaire items on desirable and probable developments are formulated. Using an online tool, the Delphi survey is conducted with a large number of experts in two separate waves. The results are then analyzed and documented.

(III) The quantitative results of the Delphi study in turn form the basis for the development of the future scenarios. The project report summarizes in synoptic fashion the experts' responses to flesh out consistent "images of the future." Based on the gathered data, the research team identifies the mediating factors with the most significant impact. Four scenarios are developed:

1. The first scenario is marked by a low level of technological development as well as greater cultural acceptance of uncertainty and risk as a core aspect of human life.
2. The second scenario combines a high level of technological advancement but a broad rejection of this trend by airport users and the general population.
3. The third scenario is characterized by a high level of technological advancement including broad acceptance of this trend (including associated restrictions to personal freedom) by airport users and the general population.
4. The fourth scenario presents a low level of technological advancement coupled with a desire for the increased deployment of security and surveillance technologies in society.

The scenarios are constructed in the form of narrative stories and embellished with illustrations. The researchers devote particular attention to ensuring that each scenario is consistent with the gathered data.

The aim of the methods combined in this example is to enable a complementary view of the subject matter, as the results stand alone in each case, but complement each other in sum. In particular, the explorative interviews with experts (phase 1) enable a qualitative understanding of the problem situation from the perspective of relevant actors. The diverse facets of the research problem thus become clear. They include the social acceptance or rejection of a technology by airport passengers; technical difficulties; potential passenger data misuse; and practical considerations, such as how poorly trained security personnel deal with passengers. By contrast, the Delphi survey (phase 2) takes a more focused view: A wide range of experts are recruited to assess aspects derived from the expert interviews (phase 1) in terms of their probability of occurrence and desirability. At this point, the research team can now undertake a statistically verifiable analysis of selected aspects of the topic area based on a broader selection of respondents. In the subsequent scenario development process (phase 3), the transformation of these results into narratives about the future aims to encourage different stakeholders to think about potential future developments, even though they cannot provide all-encompassing descriptions of the future. Questions are raised to help the study's intended audience arrive at decisions regarding the future of airport security: What can be done to circumvent or to facilitate the developments elaborated as possible futures?

From a methodological perspective, it is crucial that each scenario is based on the results of the preceding methodological steps and can be traced back to them. Thus, the intention from the outset was not to compare or validate the Delphi study and the scenario process, but rather to undertake methodological steps that build on each other and refine the quality of the data. The value of the study ultimately lies in the consilience of knowledge from disparate domains. Based on the study's findings, the researchers are able to furnish evidence-based and problem-oriented action alternatives and recommendations to the executives of the airport operator. In this way, the project enhances the knowledge base regarding the problem situation while also augmenting the scope of potential action open to airport executives.

References

Gerhold, L. (2012). Methodenkombination in der sozialwissenschaftlichen Zukunftsforschung. In R. Popp (Ed.), *Zukunft und Wissenschaft: Wege und Irrwege der Zukunftsforschung*, 159–183. Springer.

Glenn, J. C. & Gordon, T. J. (2009). Integration, Comparisons, and Frontiers of Futures Research Methods. In American Council for the United Nations (Eds.), *Futures Research Methodology: Version 3.0* (Chapter 39, 1–34).

Steinmüller, K. (1997). *Grundlagen und Methoden der Zukunftsforschung: Szenarien. Delphi. Technikvorausschau.* Sekretariat für Zukunftsforschung (SfZ). WerkstattBericht Nr. 21. Gelsenkirchen.

Further Reading

Creswell, J. & Plano Clark, V. (2017). *Designing and Conducting Mixed Methods Research.* 3rd ed. Sage.

Popper, R. (2008). How are foresight methods selected? *Foresight, 10*(6), 62–89.

Tashakkori, A. & Teddlie, C. (2010). *SAGE Handbook of Mixed Methods in Social & Behavioral Research.* Sage.

Producing Quality Research

Roman Peperhove and Tobias Bernasconi

Summary

Work in futures studies poses a special challenge for good scientific practice: its objects of research are not amenable to empirical measurement and often span multiple disciplines. How can cognitive biases in future scenarios be identified and suppressed if forecasts do not permit corroboration? How does one assess the quality of the underlying data? How are researchers supposed to make valuable statements about a future that is impossible to predict? And given the field's transdisciplinary and interdisciplinary scope, which existing concepts are appropriate, and where do new ones need introducing? Research quality grows out of the implicit demands of good scientific practice and concerns topics such as the avoidance of cognitive bias, the critical examination of underlying data, the judicious selection of experts, and clarity in the use of new terminology.

R. Peperhove (✉)
Research Forum On Public Safety and Security, Freie Universität Berlin, Berlin, Germany
e-mail: roman.peperhove@fu-berlin.de

T. Bernasconi
Faculty of Human Sciences, University of Cologne, Cologne, Germany
e-mail: tobias.bernasconi@uni-koeln.de

Essentials

Futures researchers make assessments about future situations and events. To that end, they draw on disciplinary and cognitive tools to develop conceptual models, produce primary data, gather secondary data, and incorporate outside expertise. The quality of futures research depends on the application of basic standards of good scientific practice such as transparency, solid theoretical foundations, appropriate methods, and a careful execution of the elementary skills of science. Producing quality research is all the more crucial in the case of futures studies because it cannot make use of one of the most important scientific standards of them all: empirical validation (see "Validation by Argumentation").

Obtaining quality research requires adherence to a set of guidelines for good research practice:

1. Researchers must identify and, as far as possible, eliminate *cognitive biases* in their work and communication.
2. Researchers must regard all the *data* they use as potentially problematic. They must critically examine sources of information for appropriateness and quality and either justify their use or call them into question.
3. The quality and selection of external *experts* should not be taken as given but subject to critical examination.
4. *Conceptual terminology* must be developed with care and applied consistently.

Only when these guidelines governing the "craft" of science are adhered to can one speak of quality research. In this regard, quality research is both a characteristic but also a condition of good futures studies.

Guidelines

1. *Counteract cognitive biases:*
   Producing quality research begins with the researchers themselves. Researchers must reflect on and, as far as possible, reduce potential *cognitive biases* in the selection, generation, processing, evaluation, and communication of data and information. The first step is to identify individual, internal, and external factors that could influence knowledge, forecasts, and their consequences. If necessary, researchers should identify and implement suitable countermeasures. Integrating many different perspectives is one way to weed out individual and collective biases.

2. *Make sure data and information are adequate:*
   Researchers must check and, as far as possible, ensure the adequacy of information and data, i.e. its relevance and completeness with regard to the research question(s). Moreover, they must identify missing information, justify the selected sources, and explain their method for handling information deficits.
3. *Check data for quality:*
   Researchers must also reflect critically on the distortions and limitations of the data and minimize them whenever possible. They should document internal and external factors (e.g. content requirements, lack of resources, imbalances in official statistics) that can shape data collection and analysis. With empirical work, the selection of data must always be prudent whether the number of observations is large or small, and must not be determined solely by situational constraints or expectations. Researchers should consider procedures to minimize distortions and carry out a multi-layered (i.e. interdisciplinary and intercultural) analysis of the data.
4. *Examine the quality of the literature and other sources of information:*
   With regard to literature and other sources of information for a futures research project, researchers must identify factors that may impair quality and eliminate them as far as possible. The following quality criteria should be considered: current relevance, professional value, analytical depth, independence, and transparency.
5. *Select appropriate external experts:*
   Project researchers are ultimately responsible for the preparation, collection, evaluation, and interpretation of data appropriate to the research topic. But they are also responsible for selecting which outside experts to consult. The choice of outside experts should be based on professional expertise, methodological competence, practical experience, institutional ties, and the ability to think laterally in networks. It is important that researchers be able to explain their decisions and disclose their underlying understanding of expertise. If an expert is a representative or functionary of a larger institution, researchers must ensure that he or she actually contributes quality specialist expertise (see Meuser & Nagel 2002). In general, experts should have access to specific forms of knowledge or decision-making or have demonstrated that they can make a unique contribution to the discussion. Researchers must determine experts' attitudes regarding the research question and method as well as their degree of expertise (by, say, posing questions designed to measure subjective competence) (see Häder 2002). The use of an expert matrix (see Varho & Tapio 2013) can be helpful for these purposes.

6. *Ensure the quality of terms and use them in disciplined manner:*
Researchers must carefully establish and employ the basic terminology of a project. New terms should only be introduced when indispensable. In all other cases, existing and established terms should be used. Key terms for the project, whether they have been adopted from other work or newly coined, arcane or widely known, must be carefully defined at the outset. It is important that researchers use terms consistently for the duration of the project and in the documentation and presentation of the results. The same word should be used to designate identical meanings, and different words should be used to denote different meanings.

Common Shortcomings and Pitfalls

a) *Failing to identify cognitive biases:*
Researchers may tacitly assume that they are without cognitive biases because they are in agreement on all the basic issues. They fail to understand that blanket agreement should be a warning sign and a starting point for critical reflection. As a result, they do nothing to root out cognitive bias, and misinterpretations occur. Assigning one of the team members the role of devil's advocate can help prevent similar outcomes.

b) *Failing to identify institutional constraints:*
Researchers may fail to identify institutional constraints regarding the selection of search criteria and the interpretation of results (see Ansoff & McDonnell 1990). Internal, political, or strategic parameters can, intentionally or not, shape the design and evaluation of research.

c) *Assembling a research team that is too homogeneous:*
When selecting research personnel, researchers may fail to consider that personal qualities (e.g. age, gender, culture, religion, character, etc.) can impact the evaluation of information and the attribution of meaning. Homogeneity among researchers can have a large distortive effect, especially in the context of futures studies, where empirical validation is unavailable (see Moser 2011).

d) *Selecting inappropriate sources:*
Researchers may select literature and other sources of information mainly because they are easily accessible (via Google, say). Although professional, in-depth results can be had with additional effort, researchers settle for inferior sources of dubious provenance and fail to consider relevance, recentness, and validity. And in cases where there are no alternatives to unexamined, interest-driven sources, they do not mention the resulting limitations.

e) *Opting for quantity over quality:*
 When building a data foundation, researchers may emphasize quantity over quality in order to impress funders or demonstrate expertise. As a result, they fail to assess quality and key signals get drowned out by the noise.
f) *Using poor data:*
 Researchers may use poor data and fail to identify deficiencies. They may draw the data from suboptimal sources, fail to identify relevant information, or pass up the opportunity to offer grounded hypotheses to fill the gaps.
g) *Creating an inadequate research design:*
 Researchers may collect information and carry out empirical surveys only for cases that are currently available or easily accessible. Moreover, they may fail to question whether the observed cases are relevant to the present research, or they may carry out empirical surveys solely for the sake of having empirical data and then do nothing with the results in subsequent stages of the project.
h) *Misconstruing expertise:*
 Researchers may exaggerate the significance of judgments on the basis of the fact that they stem from experts. They may assume them to be correct without critically reflecting on their contents. Regarding experts as the bearers of "true" and "correct" assessments can be counterproductive (Tetlock 2005).
i) *Poorly selecting experts:*
 Researchers may exclusively select high-ranking experts at organizations and institutions. But the most experienced experts are often found at the second or third tier, as this is where decisions are prepared and where most knowledge is available (see Meuser & Nagel 2002, p. 74).
j) *Using imprecise terminology:*
 Researchers may misunderstand work on the future as carte blanche for carelessly and needlessly introducing new terminology. The terms are meant to convey expertise but ultimately obscure rather than elucidate the facts.
k) *Poorly defining and using terms:*
 Researchers may leave central terms in their work undefined and without proper introduction. The terms may disappear just as abruptly as they appear. Researchers may use different words to describe the same phenomena for the sake of variation. It is crucial that they use terminology consistently, however.

Illustrative Example

Researchers are tasked at any early phase of a security-related project with assessing the potential misuse of technologies in the future. The challenge is to identify the technologies of tomorrow – some yet to be invented – that could be used for criminal or violent ends. First, the researchers define the tasks and individual steps: 1) identify developing technologies; 2) evaluate them for potential misuse; 3) evaluate the effects of misuse; and 4) discuss measures to prevent or impede misuse in a timely fashion.

The researchers are initially concerned that they will be told in advance what results to deliver – a phenomenon not uncommon in contract research – and if so, they vow not to continue. Fortunately, however, the client has decided to give them free rein. They quickly determine that the term "technologies" is too broad for the purposes of the research. This is a common problem with futures studies: because a great many developments are conceivable, there is the risk of getting lost in the seemingly infinite possibilities. So they consult with the client and based on the feedback they receive narrow the scope to nanotechnology and robotics.

With the subject area still broad, they employ a technique known as *horizon scanning* to gain an understanding of the most recent developments. In the internet age, gathering information is easy; some simple keywords is all it takes to find a trove of journals, blogs, and databases. They select sources based on relevance and credibility and determine whether the information they have gathered is sufficiently comprehensive. They pay special attention to the question of whether the data can deliver meaningful results in line with the research question. To make sure they have a solid foundation for their subsequent work, they ask experts from nanotechnology and robotics to examine the sources for reliability and completeness. The researchers incorporate the experts' feedback, and are careful to identify the inevitable gaps in the data (such as inaccessible sources and uncertainties in the data).

The researchers note that the homogeneity of their team – all men, roughly the same age, with similar backgrounds – could prejudice their selection and assessment. Accordingly, they engage an array of experts from diverse professions and countries of origin to evaluate their work for cognitive bias. Then, the researchers ask international technology and security experts to evaluate the results of the horizon scanning (see Douw & Vondeling 2006). The experts use a standardized catalogue of questions and criteria based on the Delphi method to assess the likelihood of market entry and misuse (based on criteria such as complexity and possible regulatory protections) as well as the possible consequences.

The researchers realize early on that their team members have a different understanding of some of the terms and objectives. So they develop a glossary for communication within the team as well as with the client. They ask that all participants use terms clearly and consistently in order to prevent ambiguity. This is a constant challenge in interdisciplinary projects, especially when they involve futures research.

Based on the Delphi evaluations, the team develops several narrative scenarios that present the results vividly and encourage the participation of additional experts. In an interdisciplinary workshop, they discuss the misuse potential of the selected technologies and represent them in short narratives. The researchers set aside two days for the workshop, as they have found that important insights often surface on the second day. The results are surprising. Not only do the participants respond positively to the scenarios; they realize that some seemingly harmless technologies could lend themselves to misuse.

By varying the factors and weightings, the researchers produce a range of scenarios, some mutually exclusive. Each is internally coherent and helps with the visualization and assessment of individual technologies. In accordance with the research objective, the team also identifies the technologies with the greatest potential for misuse and presents them by order of priority (using percentages and scales). The team orders the technologies mainly based on the Delphi results, but they also include them in the scenarios to get a sense of their effects in everyday settings. Only when the results and the scenarios are coherent do they finalize the analysis.

The researchers document the process so the client and external experts can understand their choices, ensuring transparency. In forestalling any suspicions that they were unduly influenced by institutional interests or by current political and societal debates, the researchers outline their work steps and describe the quality, diversity, and selection of research staff, experts, and data. They are keen to avoid the mistake of a previous team, who presented only their results and were immediately accused of being one-sided and unprofessional.

In sum: the project team increased the quality of their study by counteracting cognitive biases, by paying great attention to the appropriateness of their data and experts, and by carefully developing and employing their terminology.

References

Ansoff, H. I., & McDonnell, E. J. (1990). *Implanting strategic management* (vol. 2). Prentice Hall.

Douw, K., & Vondeling, H. (2006). Selection of new health technologies for assessment aimed at informing decision making: A survey among horizon scanning systems. *International Journal of Technology Assessment in Health Care, 22*(2), 177–183.

Häder, M. (2002). *Delphi Befragungen. Ein Arbeitsbuch.* Wiesbaden: Westdeutscher Verlag.

Meuser, M. & Nagel, U. (2002). Experteninterviews – vielfach erprobt, wenig bedacht. Ein Beitrag zur qualitativen Methodendiskussion. In: Bogner, A., Littig, B. & Menz, W. (Eds.) *Das Experteninterview. Theorie, Methode, Anwendung,* 71–93. Opladen, Leske + Budrich.

Moser, S. (Ed.). (2011). *Konstruktivistisch Forschen. Methodologie, Methoden, Beispiele.* Wiesbaden: VS Verlag für Sozialwissenschaften.

Tetlock, P. (2005). *Expert political judgment: How good is it? How can we know?* Princeton University Press.

Varho, V., & Tapio, P. (2013). Combining the qualitative and quantitative with the Q2 scenario technique – the case of transport and climate. *Technological Forecasting and Social Change, 80*(4), 611–630.

Further Reading

Breuer, F., & Reichertz, J. (2001). Science criteria: a moderation. *Forum Qualitative Social Research* (Online Journal), *2*(3).

Flick, U. (2009). *An introduction to qualitative research.* London: Sage.

Mayring, P. (2014). *Qualitative content analysis: theoretical foundation, basic procedures and software solution.* Klagenfurt.

Schwartz, P. (1996). *The art of the long view: Paths to strategic insight for yourself and your company.* Random House LLC.

Smith, J., Cook, A., & Packer, C. (2010). Evaluation criteria to assess the value of identification sources for horizon scanning. *International Journal of Technology Assessment in Health Care, 26*(3), 348–353.

Steinke, I. (1999). *Criteria of qualitative research: Approaches to the evaluation of qualitative-empirical social research.* Juventa.

Scientific Relevance

Birgit Weimert and Axel Zweck

Summary

Research results are scientifically relevant if they help expand the knowledge base, advance our understanding of a certain subject, or provide interdisciplinary insights. In other words, they represent findings that are novel, worth knowing, and accepted by the scientific community. Scientifically relevant information serves the interests of scientific inquiry and emerges from good scientific practices, i.e. from processes that are transparent and public.

Essentials

When it comes to futures studies, the question of "relevance" is complicated. Scientific relevance is important, but so too are practical relevance and social relevance. Practically relevant research solves applied problems, while socially relevant research concerns questions related to human coexistence. The latter two types of research can be very relevant in terms of their utility but scientifically unimportant because they contribute nothing of scientific value (see "Practical Relevance, Usefulness, and Effectiveness").

B. Weimert (✉)
Corporate Technology Foresight, Fraunhofer Institute for Technological Trend Analysis INT, Euskirchen, Germany
e-mail: birgit.weimert@int.fraunhofer.de

A. Zweck
Innovations- und Zukunftsforschung, RWTH Aachen University, Aachen, Germany
e-mail: azweck@soziologie.rwth-aachen.de; zweck@vdi.de

© The Author(s), under exclusive license to Springer Fachmedien Wiesbaden GmbH, part of Springer Nature 2022
L. Gerhold et al. (eds.), *Standards of Futures Research*, Zukunft und Forschung,
https://doi.org/10.1007/978-3-658-35806-8_12

What sets scientifically relevant research apart from other forms of relevance is that it helps answer scientific questions. It helps increase knowledge in a particular discipline or an interdisciplinary community and contributes to the existing knowledge base. Of course, the evaluation of scientific relevance always takes place in a historical context, shaped by particular social and scientific values. Accordingly, what counts as scientifically relevant can and does change.

Scientific relevance depends on the prevailing standards in the scientific community. They determine what counts as interesting, impactful, novel, original, and universal, and are closely related to issues such as theoretical models, transparency, objectivity, reliability, validity, precision, honesty, intelligibility, logical reasoning, simplicity, and equilibrium (Kreibich 2006). One common scientific standard does not apply to futures studies, however. Futures research is not verifiable, at least not in the usual sense of the term (see "Validation by Argumentation"). Because it concerns events in the future, corroborating empirical data does not exist. But this does not make it invalid. "If a certain development is postulated," Axel Zweck observes, "it is not determined a priori that the said postulate was wrong in the event the respective development does not occur" (2005, p. 11).

Scientifically relevant results are, therefore, not limited to gains in knowledge; nor do they result exclusively from procedures that adhere to the rules of the scientific community. Other important factors include the relevance of the topic, the creativity of the research design, and the contribution to scientific progress (i.e. scientific impact).

Research topics often become relevant due to social changes or disputes in the scientific community. The topics may either serve to underpin or challenge central paradigms (see Kuhn 1976). The fact that research is often concerned with practical issues does not mean that the given challenges can be solved exclusively by science.[1] At the same time, what may at first glance seem purely practical could also be scientifically relevant. In this case, the topic may require reformulating in more general terms or combining with knowledge acquired from other applications. Issues of scientific relevance and practical relevance (or societal relevance) are not antithetical but intertwined, even if economic utility and practical relevance are generally not indicators of scientific relevance (see Bender 2001; Dilger 2012).

Two crucial factors for scientific advancement are originality and creativity (see Heinze et al. 2013). This is the reason why Popper describes research as a

---

[1] The implementation of the research results must "neither be immediate, nor [must] the intended benefit exactly correspond to the actual benefit" (Baumgarth et al. 2009).

creative art (see Popper 1996). Creativity has various layers, ranging from original research questions and research design to novel interpretations of the results and innovative forms of presentation.

The scientific impact is the extent to which results and practices are accepted by the scientific community (frequently measured by the number of citations). Typically, research continues until it has a public effect (see Werth & Sedlbauer 2011). In many cases of contract research, however, secrecy, intellectual property protection, or narrow specificity preclude that outcome. Moreover, public projects can lack scientific relevance, such as those that focus exclusively on the final report. It is important, therefore, to distinguish between a project's scientific impact and the impact intended by executive boards, politicians, or the public at large. Frequently, futures research is initiated with the intention of being a marketing instrument, not a vehicle for new knowledge or a solution to real-world problems.

Another important criterion for scientific relevance is universality, i.e. the extent to which researchers aim to expand the reach of theoretical models into something more generalizable. But the process of generalizing research questions, which are often quite specific, requires a careful examination of their many interdependencies along with a firm grasp of the larger context. Researchers need to have a comprehensive understanding of future relevance, the likelihood of change, and applied potential (see Weimert 2012). An overly narrow purview at the outset not only makes it difficult to define a research question; it also reduces the quality of the results and may limit their scientific relevance.

Not all factors measuring scientific relevance can be quantified in the way that publications, citations, or funding can, and scientific relevance is often difficult to verify when it comes to futures research in particular. At the same time, science does not always adhere to stated principles of rationality. It, too, is subject to ever-changing trends and movements. In retrospect, scientific practice is often interpretable as a product of social circumstances, world views, political interests, laboratory practices, (see Knorr-Cetina 1984), and scientific paradigms (see Kuhn 1976; de Solla-Price 1974).

Guidelines

The scientific relevance of futures research mostly depends on the level to which it satisfies the following criteria:

1. *Reflection of current state of research:*
   Insights and findings must take into account existing knowledge in the area under study.
2. *Novelty:*
   Insights and findings must make a novel contribution to the area under study.
3. *Scientific progress:*
   Insights and findings must help answer open scientific questions, close gaps in scientific knowledge, or formulate new lines of questioning.
4. *Transparency:*
   Insights and findings must be derived from transparent research that is rigorously supported or carefully examined.
5. *Scientific debate and quality control:*
   Insights and findings must be subject to debate in the scientific community, which will assess its quality, reproducibility, and transparency in accordance with the current state of research.
6. *Scientific impact:*
   Researchers should take care to ensure that the results and practices are transmitted to the scientific community in the form of publications and talks.
7. *Relevance of the research problem:*
   The research problem should be viewed as urgent and relevant by the scientific community.
8. *Universality:*
   Whatever difficulties encountered in practical implementation, the objective of scientific work must be to see whether the findings and theories can be generalized.
9. *Originality and creativity:*
   Care must be taken to ensure that the contribution is scientifically original, i.e. that it opens new horizons and that the findings are non-trivial.
10. *Validity of standards:*
    Researchers must examine whether existing standards for scientific relevance are still valid for the work. This includes the assessment of research questions for relevance, their potential compatibility with ethical principles, and the social responsibility of individual scientists and the scientific community as a whole.

Common Shortcomings and Pitfalls

a) *Failing to participate in the scientific community:*
   Researchers may focus too much on the final report and/or presentation for the project initiator. They critically examine their approach and results, but forgo lectures and publications about their work because they do not serve marketing. They publish results only when opportune.
b) *Allowing results to be shaped by stakeholders:*
   When performing contract research, investigators may align the results they deem scientifically relevant with the expectations of the funder without disclosing this to the scientific community. As a consequence, external researchers are unable to fully perform their regulatory function.
c) *Providing results that are too specific:*
   Practical relevance is essential for the success of futures research. Practically relevant results *can* be scientifically relevant, but not always. In contract research, investigators may concentrate on an overly narrow question that is unable to generate results that are relevant for the scientific community.
d) *Failing to define an original research question or project:*
   Investigators may confirm prevailing axioms but fail to produce new observations pertinent to current research.
e) *Failing to validate the results sufficiently:*
   Researchers may publish their results before subjecting them to a scientific review or to critical reflection. If the results attract public attention, the reputation of science as a whole may suffer, as was the case with the "discovery" of cold fusion.
f) *Failing to be critical of scientific-seeming institutions:*
   Findings from universities and other research institutions may be more likely to be regarded as scientific than those obtained by companies or consulting firms. Private companies may seek to counteract this bias by encouraging their researchers to take teaching positions at universities or by choosing a scientific-sounding name. It is important, however, that they not attempt to "elevate their business with the aura of academic seriousness" (Rust 2012, p. 53). At the same time, scientists are increasingly under competitive pressures[2] (see Winter & Würmann 2012), which can result in abuse or a lack of rigor (see German Research Foundation 1998).

---

[2] This aspect includes project success rate and quantitative performance measurements.

Illustrative Example

The new management of a well-established tech company determines that it needs to conduct more long-term forecasting. So it forms an internal working group to consider possible transformations in society and the likely consequences of technological developments. The company aims to increase its profitability by dedicating itself to cutting-edge research. Moreover, it decides to collaborate with the scientific community for inspiration and legitimacy. Through conferences, workshops, and reviews of the scientific literature, the company's researchers assess the novelty and relevance of its work.

The company develops an approach that is tailored to its particular corporate structure and operations, attaching great importance to reliability and regularly publishing the results. In order to promote their employees' creative potential, originality, and breadth, it reconfigures its office space, creating a more open and collaborative working environment.

Insofar as trade secrets allow, the company shares its procedures, applied methods, and results with the scientific community by means of journal articles, conference talks, and book chapters. The knowledge gained from the work with the scientific community serves as an additional source of inspiration. The results are not only practically relevant for the company but scientifically relevant for the research community, garnering considerable attention from specialists.

While competitors continue to rely on practical approaches that serve their own ends without consulting the scientific community, the company expects that its new scientific bona fides will lead to better market outcomes. It publishes future scenarios at regular intervals designed to further the chances of market success. Specifically, it hopes that the resulting discussions will garner more positive media attention for its work and prepare the public for the developments envisioned in the scenarios. By impacting stakeholders in government and business, the scenarios support its corporate objectives. The scientific relevance of the scenarios increases their social relevance because the public understands that they serve the greater good.

A difficult yet crucial step in the success of the company is then to determine the best corporate and product strategies that follow on from the scenarios.

References

Baumgarth, C., Eisend, M., & Evanschitzky, H. (2009). *Empirische Mastertechniken.* Springer.

Bender, G. (2001). Einleitung. In Gerd Bender (Ed.), *Neue Formen der Wissenserzeugung*, 9–22. Frankfurt a. M.: Campus.

Dilger, A. (2012). Rigor, wissenschaftliche und praktische Relevanz. Diskussionspapier für das Institut für Organisationsökonomie, 03/2012. http://www.wiwi.uni-muenster.de/io/forschen/downloads/DPIO_03_2012.pdf. Accessed 17 May 2021.

Heinze, T., Parthey, H., Spur, G., & Wink, R. (2013). *Kreativität in der Forschung.* Wissenschaftsforschung Jahrbuch 2012. Berlin: Wissenschaftlicher Verlag Berlin.

Knorr-Cetina, K. (1984). *Die Fabrikation von Erkenntnis: Zur Anthropologie der Naturwissenschaft.* Frankfurt/Main: Suhrkamp.

Kreibich, R. (2006). *Zukunftsforschung.* IZT – Institut für Zukunftsstudien und Technologiebewertung. Arbeitsbericht Nr. 23/2006. Berlin. http://www.izt.de/fileadmin/downloads/pdf/IZT_AB23.pdf. Accessed 17 May 2021.

Kuhn, T. (1976). *Die Struktur wissenschaftlicher Revolutionen.* Frankfurt/Main: Suhrkamp.

Popper, K. (1996). *Alles Leben ist Problemlösen: Über Erkenntnis, Geschichte und Politik.* Munich: Piper.

Rust, H. (2012). Schwache Signale, Weltgeist und "Gourmet-Sex." In R. Popp (Ed.), *Zukunft und Wissenschaft: Wege und Irrwege der Zukunftsforschung*, 35–57. Springer.

Solla-Price, D. (1974). *Little Science, Big Science: Von der Studierstube zur Großforschung.* Frankfurt /Main: Suhrkamp.

Weimert, B. (2012). Der Blick auf die Technologien von morgen. *Wissenschaftsmanagement*, 4, 42–45.

Werth, L., & Sedlbauer, K. (2011). *In Forschung und Lehre professionell agieren.* Bonn: Deutscher Hochschulverband.

Winter, M., & Würmann, C. (Eds.) (2012). Wettbewerb und Hochschulen. 6. Jahrestagung der Gesellschaft für Hochschulforschung in Wittenberg 2011. *Die Hochschule, Journal für Wissenschaft und Bildung,* 2/2012.

Further Reading

German Research Foundation (1998). *Vorschläge zur Sicherung guter wissenschaftlicher Praxis.* Empfehlungen der Kommission "Selbstkontrolle der Wissenschaft." Weinheim: Wiley-VHC Verlag.

Zweck, A. (2005). Qualitätssicherung in der Zukunftsforschung. *Wissenschaftsmanagement*, 2, 7–13.

Codes of Conduct and Scientific Integrity

13

Andreas Weßner and Elmar Schüll

Summary

Trust in the integrity, honesty, and quality of research is a pillar of both the scientific community and society in general. Without a sense of what counts as an honest scientific endeavor (see SCJ, 2006, p. 5; and the National Academy of Sciences, 1995, p. 5 ff.), it would be impossible to build on other's work or apply research results with confidence. Codes of conduct and principles of good scientific practice are indispensable tools for securing trust in a researcher's work and reputation. They reflect the values that should underlie everyday scientific practice, and are guidelines on which both researchers and non-researchers alike can rely. As high-profile incidents of scientific misconduct make clear, however, researchers do not always abide by the rules. Various bodies and professional societies have published explicit codes of conducts to discourage transgressions and ensure research quality.

A. Weßner (✉)
Institute for Technology and Work (ITA) e.V., Kaiserslautern, Germany
e-mail: andreas.wessner@ita-kl.de

E. Schüll
Salzburg University of Applied Sciences, Salzburg, Austria
e-mail: elmar.schuell@fh-salzburg.ac.at

Essentials

The German Research Foundation (DFG) recommends that all scientific societies draw up standards of good scientific practice and require their members to comply with them (see DFG, 2013, p. 14). For futures research in German-speaking countries, there is as yet no corresponding code of conduct. The present text suggests some basic principles for such a code.

Futures research draws on findings from other disciplines, combines them to address cross-disciplinary questions, and uses its own methods to generate knowledge pertaining to the future (Zweck, 2012, p. 69). In the German-speaking world, at least, futures studies is still a young discipline and lacks a system of peer review (ibid., p. 73). Nevertheless, public institutions, politicians, and businesses frequently look to futures researchers for reliable information and guidance. This makes it all the more necessary to formulate principles of scientific practice for futures researchers, particularly with regard to their core tasks of providing orientation, aiding decision-making, and identifying alternative courses of action.

Codes of conduct cannot replace quality assurance bodies, and vice versa; each must complement the other to be truly effective. Cases of scientific misconduct in developed disciplines that have established control bodies show that, besides third-party monitoring, the research community needs a common set of values to emphasize the responsibility of individuals and institutions and to minimize the occurrence of intentional and unintentional misconduct. Moreover, principles of good scientific practice publicly communicate criteria by which to measure research and to which researchers commit themselves.

Principles of good scientific practice govern the chosen research design, the range of applied methods, and their discipline-specific criteria. But they also insist on the personal responsibility of researchers, serve as guidelines for self-monitoring, and define the boundaries of scientific misconduct.

Guidelines

We have drawn on the recommendations of the German Research Foundation[1] and the European Code of Conduct for Research Integrity (see ESF, 2011) to formulate the following principles of good scientific practice for futures research:

---

[1] Unless otherwise indicated, the following guidelines are based on the principles for good scientific practice recommended by the German Research Foundation (DFG 2013).

1. *Adhere to basic principles*:
   a) The researcher is responsible for the research design he or she chooses, the methods applied, and the documentation of the research results.
   b) The research design must be appropriate to the issue under investigation and aim to achieve unbiased results *lege artis*. This goes hand in hand with the task of explaining and reflecting on the project's underlying objectives, framework, and values (see "Objectives and Framework Conditions" and "Transparency").
   c) Futures researchers must explicitly state which scientific methods were applied to achieve the results and which limitations are associated with the use of those methods or with the research design in general. This especially applies when combining methodologies (see "Method Selection").
   d) Futures researchers must state explicitly and as concretely as possible the degree of uncertainty that accompanies their research (see National Committees for Research in Norway 2008, p. 8; and National Committees for Research in Norway 2009, p. 14).

   Researchers must document the research design applied and publish their findings, citing primary and secondary sources.

2. *Disclose conflicts of interest*:

   It is paramount that any potential conflicts of interest be communicated early on. Such conflicts are at odds with independent and open-ended research, and can jeopardize the quality of work and cause lasting damage to the reputation of researchers and the scientific community as a whole.
   a) Futures researchers must disclose potential conflicts of interest and, in particular, state whether and to what extent their research is funded by third parties (be it a natural person or a legal entity).
   b) It is virtually impossible to develop an original research question or to approach a research project without prior assumptions or preferences. This makes it all the more imperative that futures researchers clearly distinguish empirically based statements from their personal viewpoints.

3. *Organize work groups responsibly*:

   Working groups must be organized so that individual results can be communicated, constructively criticized, and compiled into a common body of knowledge. It is crucial that members trust one another: only then will constructive discussions and debates be possible.
   a) Each researcher is responsible for his or her work.
   b) As a rule, working groups should have a leader to coordinate cooperation, regulate conflicts, and ensure quality.

c) It is the responsibility of the working group leader or the project management to provide appropriate professional supervision for younger team members (graduate students, first-year postdocs, undergraduates, etc.).
4. *Store primary data*:
   Futures researchers must ensure that the primary data used for scientific publications are stored securely for at least 10 years.
5. *Publish scientific results*:
   a) Futures researchers are responsible for the quality of their publications.
   b) Futures researchers must completely and clearly describe the methods they used, including associated limitations, when publishing their findings. Any necessary preparatory work, whether performed by the author or by another researcher, must receive explicit mention.
   c) Futures researchers are responsible for clearly indicating which contributions are their own and which are from colleagues, competitors, or previous authors.
   d) If several authors are involved in a publication, only those who have actively contributed may be named as authors.[2] This excludes so-called honorary authorship[3] as well as authorship assigned solely due to seniority within an organization, due to a funding role, or due to collaboration in a working group or in data collection.
   e) In the case of joint publications, all authors share responsibility for the results.
6. *Be aware of individual responsibility*:
   Every futures researcher is responsible for acting with integrity and in accordance with the principles of good scientific practice. Futures researchers with a management role have the additional task of ensuring that good scientific practice is followed by those under their supervision.
   a) All persons involved in the research must be sufficiently qualified to carry out the research design or apply the chosen methods.
   b) If several scientists are involved in a research project or a publication, all are equally responsible for adhering to good scientific practice with regard to the chosen research design, the methods used, and the documentation of the results.

---

[2] Specifically, only those who have contributed significantly to the conception of the research project, to the development, analysis, and interpretation of the data, and to the writing of the manuscript can be considered as authors.

[3] Here, honorary authorship refers to the naming of an author who is an authority in the field but who has not made a significant contribution to the publication.

7. *Be aware of social responsibility*:
Research does not take place in a vacuum. This is particularly true for futures research, which is usually project-based and often related to non-academic fields. If the aim of research is to solve a practical problem, to answer a practice-relevant question, to facilitate a development, or to provide decision-making support, researchers must think about the role that they play and the ethical and social responsibility that accompanies their work. Accordingly, futures researchers must be aware of the prerequisites, circumstances, and potential consequences of their work outside the scientific community.

Common Shortcoming and Pitfalls

A. *Failing to create trust*: Working group members do not trust each other; and technical issues cannot be discussed openly and constructively. The lack of openness, especially between members and the project management, impedes the evolution of new knowledge.
B. *Using honorary authors*: Individuals are granted authorship who have not made significant contributions of their own. Such "honorary" authorships are frequently a form of deference paid to prominent researchers or to heads of research institutions on whom the researchers' careers depend.
C. *Poorly designing research*: The research design is inconsistent and the methods used are obviously inappropriate for the subject based on the current state of research.
D. *Manipulating research results*: Unwanted results are made to fit the expectations of the researcher, the working group, the funder, or existing models through deliberate misinterpretations, concealment, or the selective use of data.
E. *Failing to state prior assumptions*: Researchers do not make their prior assumptions explicit, and the entire project becomes an effort to justify initial preferences and expectations.
F. *Concealing conflicts of interest*: Researchers do not disclose potential conflicts of interest (such as funding dependencies) early on, which can threaten the quality of the results and the reputation of the research project.
G. *Agreeing to contractual restrictions*: Secrecy clauses or confidentiality agreements are a frequent reality of contract research, but they are at odds with the demand for transparency in scientific work. Although confidentiality clauses can make research possible in the first place, they undercut good scientific practice by hindering discussions and criticisms of the results.

H. *Instrumentalizing research*: The funder is interested less in an open-ended, scientific approach to the research question than in an ex-post legitimization of previous decisions as forward-looking, innovative, or without alternative.[4]

Illustrative Example

A multidisciplinary team of researchers is investigating transport strategies for a project on future drive systems and the potential efficiency of intermodal networks in 2043. Due to the broad spectrum of topics, the researchers decide to break down the work into smaller work packages: 1) technical questions on transport and drive types, 2) surveys on user attitudes and transport behavior, and 3) the macroeconomic implications of new driving technologies, behavioral changes, and intermodal transport networks. The project manager appoints a leader and a core team for each work package. The teams consist of engineers and natural scientists in work package (1), sociologists in work package (2), and researchers from a variety of disciplines in work package (3).

At the outset, the project manager puts enormous value on good scientific practice and adherence to ethical standards. In keeping with the bar set by the manager, an engineer admits a possible conflict of interest: in the past he participated in several R&D projects for an international automotive supplier. In an open discussion, the working group decides that there are no conflicts of interest in this case because the projects date back several years and no future ties to the automotive supplier are foreseeable.

Research colloquia take place regularly over the duration of the project. They give team members time to clarify technical questions and the project management opportunities to check the quality of the interim results and to consider the current state of research. Together with the team leaders, the project management ensures that team members are qualified to handle their tasks professionally and efficiently. All data and results are stored on a central platform accessible to the project researchers. The management encourages the working group leaders to provide supervision to postdocs and other young researchers by involving them in the study of complex scientific issues and assigning them a mentor. Regarding dissemination, the teams agree that only researchers who have actively contributed to an article will be named as an author. In addition, the working groups aim

---

[4] See the discussion of hidden agendas in Chap. 7 ("Objectives and Framework Conditions") of this volume.

to take part in regular exchanges of ideas such as conferences and to strive to publish in leading scientific journals. In addition to presenting results for public discussion, placement in well-regarded publications will subject them to peer reviews from other researchers.

At the end of the project, the teams compile the primary data and put it in safe keeping so that it can later be checked by third parties if needed.

References

Deutsche Forschungsgemeinschaft (DFG). (2013). *Vorschläge zur Sicherung guter wissenschaftlicher Praxis: Empfehlungen der Kommission "Selbstkontrolle der Wissenschaft"*. Wiley-VCH.
European Science Foundation (ESF), All European Academies (ALLEA). (2011). The European code of conduct for research integrity. Strasbourg.
National Committees for Research Ethics in Norway (Forskningsetiske Komiteer). (2009). Risk and uncertainty—As a research ethics challenge. National Committees for Research Ethics in Norway.
National Committees for Research Ethics in Norway (Forskningsetiske Komiteer). (2008). Guidelines for research ethics in science and technology. National Committees for Research Ethics in Norway.
Zweck, A. (2012). Gedanken zur Zukunft der Zukunftsforschung. In R. Popp (Ed.), *Zukunft und Wissenschaft*, 59–80. Springer.

Part III
Standards of Group 3: Practical Relevance and Effectiveness

Dirk Holtmannspötter, Beate Schulz-Montag, and Axel Zweck

Futures researchers are often expected to provide precise forecasts so that decisions can be made about important issues, particularly in the domains of government policy or business strategy. However, as explained in the first part of this volume ("Standards of Group 1: The Future as a Subject of Inquiry"), all forecasts eventually run up against insurmountable epistemological limits (see "Images of the Future"). Any promise to predict the future with certainty belongs to the realm of prophecy. What futures researchers can do is *foresee* developments that they deem likely from today's perspective for the purpose of guiding long-term decision-making. Careful, well-supported images of the future can expand the array of decision options available to funders and other stakeholders.

The stakeholders that make up the research audience of futures studies can include government agencies, think tanks, NGOs, private companies, and the scientific community itself. Accordingly, the concerns of stakeholders can range widely, from speculative considerations of possible futures to pressing problems in need of solutions. Those varying concerns typically require a specific *modality* of futures research (see "Modality")—in other words, are the scenarios meant to represent possible, probable, or desirable developments?

The insights and recommendations furnished by futures research can shape the future, albeit indirectly. In particular, futures research generates various forms of knowledge for decision-making, including knowledge about mediating factors and potential future interactions; knowledge concerning which goals and purposes are worth pursuing; and knowledge concerning current and anticipated problems. Of course, whatever the knowledge that futures research delivers, the future remains open and, by extension, more or less uncertain. Nevertheless, the knowledge generated by futures research helps reduce uncertainty and—no less important—identifies the degree to which it remains.

Futures research often takes the form of project-based contract research. Practical relevance and usefulness for the audience are the central criteria for judging the quality of its output. To whatever extent possible, researchers should thus work to ensure effective results that satisfy these criteria (see "Practical Relevance, Usefulness, and Effectiveness"). Science and practice are not necessarily at odds; indeed, high-quality applied research is generally attainable only if its methods are sound. What is problematic is when, due to lack of resources, time, or knowledge, researchers ignore the principles of good scientific practice to cut corners and reach objectives more quickly.

Understanding the specifics of the audience—their basic functions, their decision-making procedures, their mission, their resources—is of central importance for effective futures research. The impact of futures research varies considerably depending on whether it is aimed at government agencies or firms, or whether it is intended to contribute to societal discussions more generally. Substantial differences also arise from the stakeholders' organizational size, decision-making authority, and remit. For applied studies, researchers must factor in the interests of the specific addressees along with their societal roles and taboos while maintaining a critical perspective. Any results or recommendations that might prove politically or socially unacceptable should be discussed with the funder as early as possible (see "Understanding the Type, Role, and Specificity of the Research Audience").

Results that cannot be utilized in practice are generally of little value to the research audience. High-quality research understands the dissemination of the results to be an integral part of the process. Futures researchers should prepare and communicate their findings so that they resonate with the audience. In particular, results should be intelligible yet not oversimplified. Achieving this requires thinking about communication strategies during the planning phase. The funder should be involved in the process at an early stage to understand the possible effects of certain strategies and to aid the communication of results within stakeholder organizations (see "Transferability and Communication of Results").

Good futures researchers concede that the future is contingent and uncertain, though this does not prevent the generation of useful insights for decision-making. What is important is that researchers disclose the level of uncertainly attached to their findings. They should also identify starting points for future policies and measures. The recommendations provided should consider the resources and real-life decision-making contexts of the funder and other major stakeholders. Depending on the specific task, futures researchers should also identify constructive options for action in view of audience goals while highlighting possible consequences (see "Identifying Decision-Making Spaces and Options").

Undertaking these tasks requires futures researchers to be effective project managers. They must agree on and adhere to realistic deadlines, and they must define processes, assign responsibilities, identify contacts, and develop output formats in a timely fashion. Finally, they must ensure that the project management is transparent and focused for the duration of the project (see "Project and Process Management").

Practical Relevance, Usefulness, and Effectiveness

14

Edgar Göll

Summary

Futures research always involves a host of individual and collective actors, each with their own expectations and needs (see "Understanding the Type, Role, and Specificity of the Research Audience"). For instance, in contract research funders put a premium on practical relevance and utility. Generally, future researchers should work to ensure that their results are impactful by considering possible follow-on measures as they design the study (see "Aligning Research with Ambitions for Action"). Ultimately, however, it is the findings that determine whether the research objectives have actually been achieved. Through the application of targeted models ("Theoretical Foundation"), concepts ("Operational Quality"), and methods ("Method Selection"), future researchers can produce well-founded results that add to the knowledge base ("Scientific Relevance"). Yet findings are practically relevant, useful, and effective only if they meet the knowledge requirements of the funder and are utilizable by all the pertinent actors.

Essentials

It is a basic fact of the human condition that all our thinking, decisions, actions, needs, and interests are explicitly or implicitly, consciously or unconsciously, oriented to the future. Since the mid-twentieth century, futures studies have been part of society's quest for self-awareness and reflection. One of the main objectives

E. Göll (✉)
Institute for Future Studies and Technology Assessment (IZT), Berlin, Germany
e-mail: e.goell@izt.de

© The Author(s), under exclusive license to Springer Fachmedien Wiesbaden GmbH, part of Springer Nature 2022
L. Gerhold et al. (eds.), *Standards of Futures Research*, Zukunft und Forschung, https://doi.org/10.1007/978-3-658-35806-8_14

of futures research is to improve society by providing considered, well-founded scenarios for decision-making. Practical relevance, usefulness, and effectiveness are, therefore, essential attributes of effective futures research.

Typically, the actors who consult futures research will evaluate research pragmatically based on whether it meets their requirements. Futures researchers should regularly ask themselves whether the results will stand up to such scrutiny, but it is also important that they not limit their work to actor expectations.

In testing the applicability of a study, the following questions can be useful: Do the content and time horizon accord with the realities of the funder and the other actors? How will researchers respond to the questions that are likely to be posed by the research audience? Have researchers taken into account all the factors that are essential from the actors' point of view?

The utility of futures research depends on delivering clearly formulated and easy-to-understand results by the prearranged deadline (see "Transferability and Communication of Results"). Furthermore, the results must be specific enough regarding policy decisions that it is possible to identify alternatives.

Among the most important criteria for effective futures research is that the results be reliable, plausible, and comprehensible to the actors (see "Validation by Argumentation"). Actors must find the results credible if they are to base their decisions and actions on them. Moreover, futures studies can guide decisions and actions by drawing plausible and easy-to-comprehend conclusions from the results and identifying trajectories for policy action (see "Identifying Decision-Making Spaces and Options").

Thinking about practical relevance is required not only from futures researchers but also from project partners and the involved actors, who can help identify influential factors in society, communities, and organizations and pinpoint internal conditions, traditions, experiences, interests, effects, and desires. These additional factors will aid in determining possible unintended consequences often overlooked by those who cause them. In this regard, futures research bears a special responsibility, similar to that of impact assessment for technology, legislation, and the environment.

Futures researchers should stress that the object of their work has yet to occur and depends strongly on the decisions of actors. From the outset, they should investigate the relationship to research subjects, practical relevance, usefulness, and effectiveness. This requires expertise in the conditions and dynamics of social change and in the topics, practical fields, subject areas, and disciplines along with their respective logics, modes, patterns of policy design, and processes.

Guidelines

1. *Produce results relevant to the research question*:
   The design and results of a futures studies project must hew as closely as possible to the research question. This applies to basic research as well as to contract research. The results must be plausible and researchers must account for any failures to achieve the study's objectives.
2. *Ensure practical relevance*:
   The research design must continually take into account the actors for whom the research project has or could have practical relevance beyond the funder and any affected stakeholders and participants. In addition, it must consider the range of relevant topics and the appropriate time horizon (see "Aligning Research with Ambitions for Action"). Of immediate importance are the explicit ideas and expectations raised by the funder and the other participants. It is important to come to an agreement with them about the objectives of the research and to identify any unspoken expectations. Only in this way can the funder gain a realistic understanding of the project's likely results and the project team accurately gauge the funder's position. If necessary, they should also reach an understanding on which topics, factors, and actors are relevant to the study. If they are to have practical relevance, the research results must reflect preliminary considerations and fulfil the attendant requirements. Any results that deviate from the preliminary considerations must be disclosed and clearly explained.
3. *Identify benefits*:
   Researchers must identify the possible benefits, along with their likely beneficiaries, mechanisms, and time horizon. They must analyze potential benefits not only for the economy (at the macro and business levels), but also for society, politics, ecology, and culture, to name just a few examples. (See "Identifying Decision-Making Spaces and Options.") Here again, different perspectives must be taken into account, because precise criteria do not exist for many of these utilities. In addition, researchers must ensure the practical usefulness of their work. It is imperative that the study provide insights such as surprising, new images of the future or more nuanced or better argued versions of familiar future scenarios.
4. *Meet defined objectives*:
   Researchers must consider whether the project's findings, insights, and conclusions will be effective in achieving its objectives. Of immediate importance here are the explicit ideas and expectations of the funder and participants. It is important that they agree on the underlying intentions of the projects and

discuss previously unspoken expectations (see "Understanding the Type, Role, and Specificity of the Research Audience" and "Transferability and Communication of Results"). In some situations, the effectiveness of the project, the findings, and the policy recommendations may also be relevant and should be considered as well. Researchers alone cannot guarantee the effectiveness of futures research, but they can substantially contribute to it. They must formulate the results clearly and concisely, express them in a way that the target audience will understand, draw rigorous and transparent conclusions based on explicit premises, and deliver them in advance of any important decision-making milestones.

5. *Assess unintended consequences*:
Assessing what the sociologist Robert K. Merton called "unintended consequences" is important but difficult. Clients, funding agencies, research teams, and experts typically have fairly selective ideas, ways of thinking, and work routines that bring with them certain priorities and preferences. If futures research is to be effective, it is important that researchers take these limitations into account and, if possible, overcome them. This approach will improve the quality and reliability of the results and initiate reflection on the underlying intentions.

Common Shortcomings and Pitfalls

A. *Failing to do justice to the complexity of the subject matter or to justify the focus:*
Whether due to limited funding, personnel, or time, researchers may select concepts, methods, or other tools that do not do justice to the complexity of the issues under investigation. Or they may focus on aspects, factors, and actors without providing sufficient or plausible justification. They can also err by including too many or too few of these elements, or simply pick the wrong ones. Finally, their results may over- or undershoot the targeted time horizon.

B. *Failing to clarify essential aspects of the research:*
Researchers may put aside important questions to accommodate funders, participants, or other scientists without a real understanding about which aspects, factors, and actors are essential.

C. *Failing to consider certain stakeholders*:
Habits of thinking and everyday routines can lead researchers to overlook certain stakeholders, such as those that stand to lose from a future development or policy. As a result, researchers can come to an inadequate or erroneous

assessment of the situation, and fail to include relevant concerns, interests, and alternatives in the scenarios and policy recommendations.
D. *Failing to produce new, relevant insights*:
Studies may fail to produce new insights and thus may offer only generalities or abstractions without practically relevance. In order to cover up these shortcomings, researchers may use unclear formulations and arguments or resort to technical jargon.
E. *Ignoring effects that are difficult to assess*:
When grappling with unintended consequences, researchers may fail to account for unusual, difficult-to-estimate, or long-term effects. They may meet the expectations of the funder but they pass up on the opportunity for critical reflection.

Illustrative Example

Researchers were commissioned to study sustainable development and strategy for a German state government. The responsible ministry was more interested than other governmental departments in the work, but had some misgiving about its scope. It wanted the project to study structural changes more broadly instead of focusing on individual areas. Nevertheless, all major actors of the federal state were interested in policy recommendations or had at least signaled their openness, ensuring the practical relevance of the research. Together with the funder, the research team decided to conduct an explorative study of sustainability in science, business, and water management. The team organized future workshops to examine each field and its current policies, to project future developments, and to determine which area was likely to receive the most political willpower. The team also conducted research by participating in activities within the sustainability scene. They quickly noted a disproportionately higher level of interest outside of the government, especially among civil society organizations.

In an early phase of the work, they identified the stakeholders and experts active in the three areas of activity along with the project's potential relevance and usefulness. Based on these considerations, they then invited a broad selection of actors to attend the workshops. It became clear at those gatherings how important the preliminary analysis and exchange had been for understanding the concerns and the study's potential utility—for the researchers as well as the stakeholders. Some of the workshop participants met there in person for the first time, and many had not been aware of the others' work, even though they were active in the same areas. The workshops created an intensive and constructive atmosphere, and a number of participants remained in contact and established subsequent collaborations. The participants helped actively ensure the usefulness of the project, each in a specific way.

Based on the workshops and several additional interviews, the researchers drafted a sustainability strategy for the state government. The core part of the draft consisted of ten policy recommendations, spanning a variety of actors, time periods, and levels. Thanks to the preceding work, researchers reached a realistic assessment of the situation, the intentions and potentials of the stakeholders, and the possibilities and requirements of policy action. Within only a few years, more than half of the policy recommendations developed within the framework of the project were realized. In all likelihood, this was due less to the written recommendations than to the study's agreement with actors' goals and readiness. Early actor assessment and the participatory approach resulted in a high degree of indirect effectiveness.

Further Reading

Bell, W. (2003). *Foundations of futures studies* (vol. 1). Transaction Publishers.
Bortz, J., & Döring, N. (1995). *Forschungsmethoden und Evaluation für Sozialwissenschaftler* (chap. 3, 95–126). Springer.
Bourdieu, P. (1993). *Soziologische Fragen*. Edition Suhrkamp.
Grunwald, A. (2009). Wovon ist die Zukunftsforschung eine Wissenschaft? In R. Popp & E. Schüll (Eds.), *Zukunftsforschung und Zukunftsgestaltung: Beiträge aus Wissenschaft und Praxis*, 25–35. Springer.
Kreibich, R. (2008). *Zukunftsforschung für die gesellschaftliche Praxis*. IZT—Institut für Zukunftsstudien und Technologiebewertung. ArbeitsBericht no. 29. Berlin.
Martino, J. P. (1983). *Technological forecasting for decision making* (chap. 17–19, 226–283). North-Holland.
Neuhaus, C. (2006). *Zukunft im Management: Orientierungen für das Management von Ungewissheit in strategischen Prozessen*. Carl Auer.
Rust, H. (2008). *Zukunftsillusionen: Kritik der Trendforschung*. VS Verlag.

Understanding the Type, Role, and Specificity of the Research Audience

Edgar Göll

Summary

Like modern societies themselves, the audiences for futures research are incredibly diverse. The research audience is a constitutive element of futures studies, and often figures explicitly in a project's focus. Indeed, effective futures research takes into account the constitution and interests of its audience. Doing so requires a precise analysis of the specific context and characteristics of the audience and a consideration of the ways those aspects can change over the course of a project.

Essentials

The primary audiences of future studies projects are their funders. However, futures research can directly or indirectly affect other actors as well, and may have to factor them into the project from the outset, depending on the research question.

A careful consideration of the research audience is indispensable because values always guide human action, whether implicitly or explicitly, and this applies to futures researchers as well as to those who read their studies. Accordingly, researchers must foreground their values to the extent they are known and take them into account as they carry out the project.

E. Göll (✉)
Institute for Future Studies and Technology Assessment (IZT), Berlin, Germany
e-mail: e.goell@izt.de

For example, audiences can vary greatly depending on whether they operate in politics or civil society, in the public or private sector, or at the local or international level. Each derives its legitimacy in different ways, each has different roles, and each draws on different traditions and values (see Popp & Zweck, 2013; and Göll, 2009). The different scopes of action for audiences are of particular importance for the tasks and effectiveness of futures studies.

The particular research audience will shape a project's spectrum of topics, models, and methods. For example, an audience can influence the extent to which transparency, participation, or publication is desired or intended. Even within the spheres of government or politics, audiences differ significantly in terms of whether they want the results disseminated widely or kept confidential.

Other distinguishing factors for research audiences are size, decision-making authority, area of responsibility, current situation, and mentality. For instance, large institutions are very different from small ones in terms of division of labor, staffing, and associated expertise. Situational circumstances can also play a role. Is an election campaign, a corporate restructuring, or a change in leadership in the offing? Are there ongoing scandals or political pressures, or has a particular issue captured the public imagination?

Effective futures research is based on a holistic, historically grounded understanding of social change and a close analysis of the agents who contribute to it. Hence, futures researchers must have a firm handle on the specifics of their primary audiences—their functional and decision-making logics, their hierarchies, their definitions of purpose, and their resources for action (Rogall, 2003; and Wright, 2010). At the outset of the project and throughout its course, they must reflect on the primary audience of the study and its findings (See "Practical Relevance, Usefulness, and Effectiveness" and "Transferability and Communication of Results.").

Awareness of the audience is particularly relevant for projects with an explicit practical purpose. Researchers must take into account the interests and motivations of those affected by the project directly (and, if possible, indirectly) and must incorporate their considerations into both the research design and the recommendations for action (see "Identifying Decision-Making Spaces and Options" as well as Popp & Zweck, 2013). It is important that researchers note and take seriously the specific organizational and sectoral cultures of their audience. They should bear in mind the political, cultural, or internal taboos of their audience without uncritically submitting to them.

Guidelines

1. *Consider the characteristics of the audience*:
   Because the audiences of future studies, like all social actors, are elements in and/or subjects of societal change, it is of utmost importance that they be considered in the research process. Various audience characteristics must be taken into account depending on the research question and circumstances: function, interests, resources, context, situational factors, etc. From a practical point of view, it is often essential to understand actors' decision-making authority.
2. *Understand how the audience regards the problem*:
   At the beginning of a project, researchers must determine the problems and challenges its audience seeks to address, along with the desired forms and intensity of research. Based on such considerations, they can deduce which problems and challenges to incorporate into the project and the best manner of approach, both conceptually and methodologically.
3. *Determine the forms of participation*:
   Futures research is meant to guide practical applications and assist decision-making about the future. Accordingly, researchers must early on clarify issues such as participation, mobilization, innovation, resilience, and adaptation (to new technological developments or the consequences of climate change, say). Researchers should involve stakeholders with decision-making authority in the project to improve the quality of the results, increase acceptance, and motivate action.
4. *Consider the relevant actors*:
   Depending on the research problem, the project's audience can include more than the explicitly mentioned actors. Because the unintended effects are often unclear, the circle of affected groups may be wider than it first appears. Identifying all the affected groups is difficult and time-consuming, and always involves some degree of speculation. Habits of thought and other routines can impede their identification. (Gender bias is one factor that can play role.) But a minimal clarification of affected groups with regard to their temporal, spatial, and social aspects is nevertheless necessary. It can be helpful to include unusual perspectives and actors in the project such as contrarian thinkers and lay experts.
5. *Understand the perspectives of the audience*:
   Researchers should consider their audience in terms of its size, power structure, decision-making authority, areas of responsibility, characteristics, purpose, and resources. Determining these factors will require accessing different perspectives inside and outside an organization. Researchers must make

clear who their audience is at the outset of a project and reflect on it throughout. In projects focused on real-world applications, researchers should explicitly factor the interests, motives, logics, and situational factors of the specific audiences into the research design and recommendations.

6. *Determine the power and influence of the audience*:
Determining which actors have power and the capacity to act is important for defining recommendations. They can be explicitly identified or part of larger scenarios. Researchers must pay special attention to the possibility of "veto power": Which actors could potentially block envisioned measures through legal, economic, or political means?

7. *Ensure the acceptance of research results*:
If possible, discrepancies between plausible, well-founded research and what is politically or socially acceptable, whether foreseeable or merely latent, should be clarified with the funders or the primary audience early on.

Common Shortcomings and Pitfalls

A. *Failing to gather informal, non-public information*:
Researchers may have access only to the obvious, publicly accessible facts and information about the project's audience; informal relationships and hidden agendas remain withheld from them. This leads to erroneous assessments about important variables or to conflicts regarding neglected aspects and actors.

B. *Failing to include other perspectives*:
Due to financial restrictions, time limitations, and other factors such as the role of the funder, researchers may lack the confidence and authority to factor in the parts of the audience who are critical or skeptical. As a result, they may fail to account for other perspectives and make erroneous assessments.

C. *Failing to attend to all the relevant actors*:
Due to the special effort involved, researchers may fail to consider relevant actors beyond the explicit audience. They may fail to see an actor as relevant due to habits of mind and routines ingrained in project teams ("epistemological communities").

D. *Failing to identify the intended audience with the funder*
Researchers may fail to identify the intended audience with the funder and to stress the need for a clear position at the outset of the project. The failure will have considerable repercussions on the research design and the results, and hence cannot be corrected at a later stage.
E. *Failing to involve the funder in the project*
Researchers may fail to help funders understand the research and involve them in the process. As a result, funders may not understand or accept the findings and will not be able to base their decisions on them.

Illustrative Example

Researchers are conducting a project on the experiences of the German Bundestag in addressing a key future issue and the possibilities for shaping policy around it. In a first step, the researchers interview members of a parliamentary committee devoted to the issue. In the second step, they interview representatives from other subsystems of parliament (e.g. committees, parliamentary groups). They discover that the definition and prioritization of the issue varies greatly depending on the party and, in particular, whether the party is part of the governing coalition or the opposition. For example, some parliamentary parties lack personnel resources to address the issue, and their representatives are isolated in the committee tasked with addressing it. The researchers also find that priorities among the members of specialized committees—e.g. individual MPs and staff members—differ depending on their work plans and career paths. In a third phase, the researchers interview external experts and discuss historical and international perspectives in comparison. In addition, they notice that numerous social organizations and other political actors are affected by the parliamentary activities. This impact would have gone overlooked had they focused exclusively on internal parliamentary relations. Here, the integration of external and critical viewpoints is essential.

It becomes clear to the researchers that even in the relatively straightforward structures of the German Bundestag, there is a considerable amount of internal complexity, with a diversity of concerns, traditions, interests, perspectives, motives, priorities, and resources.

For futures studies in this field of action, it is important that researchers consider the entire system of relevant committees, informal processes, and external relations to party organizations, government, media, etc. Only in this way will it be possible to grasp the positions and contexts of parliamentary subsystems and to understand the system in its complex network of relationships, dynamics, and changes. A comprehensive understanding of the audience is required for generating and formulating appropriate research results and for presenting the recommendations derived from them.

References

Göll, E. (2009). Zukunftsforschung und -gestaltung: Anmerkungen aus interkultureller Perspektive. In R. Popp & E. Schüll (Eds.), *Zukunftsforschung und Zukunftsgestaltung: Beiträge aus Wissenschaft und Praxis*, 343–355. Springer.
Popp, R., & Zweck, A. (Eds.). (2013). *Zukunftsforschung im Praxistest*. Springer.
Rogall, H. (2003). *Akteure der nachhaltigen Entwicklung: Der ökologische Reformstau und seine Gründe*. Oekom.
Wright, E. O. (2010). *Envisioning real utopias*. Verso.

Further Reading

Beck, U., Giddens, A., & Lash, C. (1996). *Reflexive Modernisierung*. Suhrkamp.
Bourdieu, P. (1993). *Soziologische Fragen*. Edition Suhrkamp.
Kristof, K. (2010). *Models of change: Einführung und Verbreitung sozialer Innovationen und gesellschaftlicher Veränderungen in transdisziplinärer Perspektive*. Oekom.
Mills, C. W. (1967). *The sociological imagination*. Oxford University Press.
Rust, H. (2008). *Zukunftsillusionen: Kritik der Trendforschung*. VS Verlag.
Treibel, A. (2004). *Einführung in soziologische Theorien der Gegenwart*. VS Verlag.

Transferability and Communication of Results

Beate Schulz-Montag

Summary

Futures research aims to address a particular audience and shape practical decision-making. This has consequences for the form of a study's final product, for the complexity of its content, and for its language. If the findings are to be effective, they must be prepared in such a way as to be easily comprehensible without oversimplifying the material. Researchers must work to encourage the implementation of the results while encouraging their critical examination. Regardless of the study's final form, however, the sharing of results will be more successful if communication strategies have been factored into the research design and if the audience can provide input while the research is still in progress.

Essentials

In all areas of inquiry that aim to support planning and action for the future, the dissemination of research results is of paramount importance. Indeed, researchers must give due consideration to developing an effective communications strategy given today's rapidly changing knowledge base, strong international competition, and complex problems affecting the economy, environment, and society. This is true not only for futures studies, but also for basic research, which depends on dialog with other areas of academia and society.

B. Schulz-Montag (✉)
Foresightlab, Berlin, Germany
e-mail: schulz-montag@foresightlab.de

Sharing the results of futures research consists of three phases, regardless of the size and scope of the study. In the first phase, researchers plan and initiate the dissemination of knowledge. This includes determining the audience, defining their participation, and planning appropriate formats to present the results as comprehensibly and engagingly as possible. In the second phase, researchers present their results and open them to critical discussion. In the third phase, researchers review and potentially hone their findings based on dialog with target audiences. Also in this stage, actors put the study's recommendations into practice. The third phase of knowledge dissemination—which includes the incorporation of research results in, for example, corporate planning, political decision-making, or other studies—is not the subject of this article, however. (For more on how futures studies can support the implementation of findings, see "Identifying Decision-Making Spaces and Options.") This chapter focuses instead on the first two phases, both of which are basic requirements for successful dissemination. They can be summed up as "ensuring the transferability of results" (phase 1) and "communicating results" (phase 2). Indeed, the success of futures research hinges decisively on findings being understood and accepted by their intended audience.

Guidelines

1. *Identify the audience early on and determine requirements together*:
   Projects have the greatest impact when the audience – the stakeholders, funders, or users – can identify with both the process and the results. Accordingly, researchers must consider how they will share the results in the planning phase and integrate these considerations into the project design. It is essential for researchers to come to an agreement with the funder on which interests are to be pursued and which characteristics the results should have (e.g. level of detail, informative value, target groups). In order to prevent misunderstanding, researchers should seek to identify potentially unstated expectations. Only in this way can the funder develop a realistic sense of the potential of options for knowledge-sharing, and take preparatory steps to support the communication of results. It is desirable, although not always possible, to involve the main actors in the research. At any rate, the dissemination of findings is most successful when there is clarity regarding intended audiences (see "Understanding Type, Role, and Specificity of the Research Audience").
2. *Select appropriate formats for knowledge sharing*:
   To share knowledge successfully, researchers must contend with numerous questions: Which presentation formats should be considered (e.g. a detailed

written report, PPT presentation, or brochure)? Should different formats be used for different audiences? When should they be made available? What formal, linguistic, and aesthetic considerations apply? Which results should be presented, and how (in a workshop, at a conference, in the media, in online forums, etc.)? Not everything that is possible and desirable is financially feasible. And not everything that the budget would allow is appropriate for the research purpose and audience. In this way, researchers should carefully consider which knowledge-sharing forms might be suitable for the project, coordinate these with the funder at an early stage, and plan resources accordingly.

3. *Present precise, easy-to-understand results*:
Both the results of the study and the approach taken in the project should be presented in a way that is understandable and clear to the intended audience, without oversimplifying the material. The content and the arguments should be presented in a transparent, comprehensible form, especially when it comes to evaluative formulations or assertions. In lieu of universally valid standards, the motto for presenting the research should be: "As nuanced as necessary, but as simple as possible." Accordingly, it might be useful to compile several versions of the presentation, each targeted to a different audience subgroup. The challenge lies in making as few concessions as possible and in following scientific protocols. For example, when comparing graphs, it is important to ensure that all scales have the same start and end points and that the cited data are accompanied by references.

4. *Craft an engaging presentation*:
Clear and engaging presentations have a much higher chance of influencing policy decisions. Yet such a presentation needs not be elaborate or cost-intensive. Accurate language, clear structure, and an aesthetically pleasing design are all that is required. Whenever possible, researchers should invest in a professional graphic appearance (yet while avoiding gimmicky or overly elaborate packaging). Because no universal rules exist for how to present research findings, researchers should orient themselves to the common communication practices and expectations of their audience. The results will be particularly memorable if the exposition is accompanied by easy-to-understand visual aids. As an added benefit, a carefully considered presentation will garner esteem and trust, thus facilitating communication.

5. *Enable effective communication*:
The communication of results is part of the research process. Furthermore, communication involves more than the mere sharing of information. Deriving from the Latin *communio*, or commonality, communication is a social action

that engenders mutual understanding, thus allowing obstacles to be overcome. In line with this insight, researchers should strive to convey findings in an audience-appropriate manner. Furthermore, researchers should encourage and moderate discussion between participants. The goal of such discussions should be planned in advance and kept in mind throughout the presentation. Should the exchange of ideas be non-binding? Should issues that go beyond the project be addressed? Should participants work to build consensus and arrive at concrete decisions on next steps? The answers to these questions should inform the presentation strategy and the choice of facilitation methods.

6. *Provide further guidance for implementation*:
To ensure the results are impactful, futures researchers should provide suggestions for implementation and subsequent activities. In certain cases, researchers may also support the implementation of the results. In any event, a quality research team should always point out the next steps, highlight the advantages and disadvantages of various options, and signal willingness to support implementation (see "Identifying Decision-Making Spaces and Options").

Common Shortcomings and Pitfalls

a) *Failing to clarify the basic parameters at the start*:
Researchers may begin their work without an adequate sense of their audience, purpose, and how they intend to present their results.

b) *Incorrect planning of time or budget*:
A fixation on substantive issues may lead researchers to overlook communication activities, thus leading to the inadequate allocation of time and resources to this important topic.

c) *Underestimating possible project consequences*:
Researchers may devote too little attention to the project's consequences (whether intended or not). Even though they may be formally responsible only to the funder, a whole range of other stakeholders may be affected by the project results. In preparing for the later dissemination of knowledge, researchers consider from the outset all those impacted, directly and indirectly. If necessary, efforts should be made to involve other stakeholders in the process.

16 Transferability and Communication of Results

d) *Failing to consider the unintended consequences of presentations*:
Example 1: The project's stakeholders may include individuals or institutions who are not present in the room when the results are presented or who are not among the primary research audience. Results that could be further disseminated by the first-order addressees must be considered with regard to the possible effects on second- and third-order recipients. For example, undesirable interactions between groups may occur if exclusive information is presented or results are reinterpreted when communicated to third parties. Example 2: Researchers may not be able to immediately answer likely questions from their audience during the presentation. This may lead individuals to question the credibility of the results as a whole, even if the researchers believe they are well-founded.

e) *Employing unsuitable presentation formats*:
A department involved in business operations may need a detailed presentation of certain facts, whereas a busy executive board may only be able to absorb a condensed overview of the material. Futures researchers must not only attend to the work but also prepare it in ways appropriate to the audience and situation. Futures studies benefit from engaging forms of presentation that enable real interaction and learning. If one presentation follows another without real discussion, such that participants return home with bloodshot eyes, a conference may fail to achieve these goals.

f) *Selecting an inappropriate event setting*:
The setting of the event to present and discuss the results may fail to meet expectations. This may result from a misjudgment of time (usually too little), the wrong location (uninviting, too small, too dark), or moderation inappropriate to the audience size.

g) *Failing to present findings in a timely manner or by appropriate individuals*:
A significant delay in communicating results may cause stakeholders to lose interest in the study or its recommendations. Furthermore, when the researcher fails to consider or actively steer how results are presented, this may lead findings to be misinterpreted.

h) *Failing to indicate possible consequences and policy options*:
Researchers may present forecasts, alternative scenarios, or other results without drawing attention to decision-making options or policy alternatives. Similarly, they may fail to provide adequate answers when individuals from the intended audience ask, "What does this mean for me?" Succumbing to this pitfall may prevent research results from being translated into concrete actions (see "Identifying Decision-Making Spaces and Options").

Illustrative Example

The governing coalition of a German federal state decides to commission a study on its future in 2030. The heads of the two parties' parliamentary groups propose the project to the state chancellery, and the government subsequently publishes a public tender specifying the main objectives. The purpose of the study is to enhance the basis for discussion within the public administration and to support the strategic planning of government policy. The public tender also indicates how the scenarios should be developed. First, the project researchers are to consult stakeholders from business, academia, and society on key future trends and challenges. Then, together with the state ministries, they are to develop alternative visions of the future and identify associated opportunities, risks, and policy options. With a view to modality (see "Modality"), the tender clearly stipulates that the future scenarios are to be explorative. They are also to take into account existing values, preferences, conditions. The primary purpose of the study is to facilitate the imagination of possible futures.

The government awards the contract to a team of futures researchers. In the preliminary discussions, it emerges that the scenarios have an indirect purpose as well: to help improve cooperation between the two governing parties and create a better atmosphere in the coalition. In talks with the government, officials introduce the idea of citizen participation. Strengthening civil society involvement has long been a concern of state politicians, who have been looking for solutions to the state's budgetary problems.

At the outset of the project, the project team plans the following forms of knowledge dissemination:

1. a detailed report that introduces the scenarios, analyses key factors, and presents policy recommendations;
2. a succinct PowerPoint presentation that contains roadmaps and conclusions for cabinet discussions;
3. an online forum where citizens can comment on the scenarios and provide input;
4. an ideas workshop for young people who, supported by artists and moderators, are to spend a day designing concrete projects for the future; and
5. a brochure summarizing the results of the scenario project for the public, to be distributed at a final press conference.

The researchers opt to present the scenarios in a narrative style, complemented by visualizations of the findings. They devote considerable attention to aesthetic and affective elements, in order to make the study clear and compelling. In addition to objective facts, the scenarios include subjective evaluations foregrounding normative decisions. In order to further enhance the effect of the scenarios, the researchers supplement some sections with real-world portrayals.

From the outset, it is clear that the development of the scenarios is to be participatory. The researchers plan several workshops, to be attended by representatives from all ministries. The report by the research team is subjected to a multi-stage revision process in which each ministry has an opportunity to comment.

The project ends with a press conference held by the president of state parliament, who emphasizes the value of the study for future policy planning.

Further Reading

Bonsen, M. (1994). *Führen mit Visionen: Der Weg zum ganzheitlichen Management*. Gabler.
Krallmann, D., & Ziemann, A. (2001). *Grundkurs Kommunikationswissenschaft*. Fink.
Krogh, G., & Köhne, M. (1998). Der Wissenstransfer in Unternehmen: Phasen des Wissenstransfers und wichtige Einflussfaktoren. *Die Unternehmung, 5*, 235–263.
Martino, J. P. (1983). *Technological forecasting for decision making* (Chap. 19: "Presenting the forecast"). Elsevier.
Steinmüller, K., & Schulz-Montag, B. (2005). *Z_Szenarioprozesse: Aus Trends und Zukunftsbildern Strategien für morgen und übermorgen entwickeln*. Z_punkt GmbH.
Stenschke, O., & Wichter, S. (Eds.). (2009). *Wissenstransfer und Diskurs*. Peter Lang.
Ungeheuer, G. (1987). *Kommunikationstheoretische Schriften I: Sprechen, Mitteilen, Verstehen*. Rader.
Wissenschaft im Dialog (WiD), Universität Stuttgart, ZIRN. (2011). *Abschlussbericht Forschungsprojekt "Wissenschaft debattieren!"* Berlin. https://www.wissenschaft-im-dialog.de/fileadmin/user_upload/Projekte/Wissenschaft_debattieren/Dokumente/Abschlussbericht_Wissenschaft_debattiert-Finalweb.pdf. Accessed 13 December 2021.

Identifying Decision-Making Spaces and Options

Klaus Heinzelbecker

Summary

In general, futures researchers identify decision-making options based on the results of their work. Researchers who recommend concrete measures and implementation procedures perform a specific type of consulting typically sought after by corporations, policy think tanks, NGOs, and the like. The boundaries between neutral research and goal-driven consulting are fluid. For each project, researchers must define them based on which future developments the funder deems possible, probable, and/or desirable (see "Modality").

Essentials

Before identifying decision-making options, researchers must undertake a thoroughgoing and careful analysis of their findings. Of course, some funders want specific policy recommendations right away, but complying with their requests can also lead to incorrect interpretations, premature decisions, and misapplied measures. To ensure that a futures study is effective, funders and researchers must agree on decision-making spaces before the contract is awarded (see "Practical Relevance, Usefulness, and Effectiveness" and "Understanding the Type, Role, and Specificity of the Research Audience"). To that end, it can be useful to sound out a range of possibilities for the objectives in question.

K. Heinzelbecker (✉)
Institut für Sales und Marketing Automation IFSMA, Ludwigshafen, Germany
e-mail: heinzelbecker@ifsma.de

The most important spaces of corporate work in futures studies are:

- the development of future markets for new products, services, or business models;
- the reinforcement of market positions or market shares for existing products;
- the evaluation of opportunities for new products to establish themselves on the market;
- the realignment of corporate strategies to take advantage of future business opportunities;
- the assessment and management of future risks in existing business areas; and
- the design of corporate structures, processes, and instruments for the future.

The above areas also apply, *mutatis mutandis*, to non-commercial organizations.

The specific decision-making spaces for policy-focused research depend on the particular government institution and its function. They can involve the future direction of economic and financial policy, the promotion of research, the development of general education, energy and environmental policy, infrastructure policy, and many other areas of policy-making.

Obviously, futures researchers must be familiar with the space of work for a given application to provide savvy decision-making options. In view of the wide range of areas for decision-making, it makes sense for futures researchers to specialize. Larger institutes and collaborative approaches (see "Interdisciplinarity") can increase the breadth of expertise that researchers can offer. At any rate, methodological competence alone is not enough for effective futures research; subject-matter knowledge and experience are indispensable as well (see "Understanding the Type, Role, and Specificity of the Research Audience").

Guidelines

1. *Tailor decision-making options to each actor group*:
   A project's spaces of work are largely determined by the remit of the funder. Typically, a company's senior management will require a broader scope of investigation than a specific division within it. The same goes for funders in the policy sphere or the non-commercial sector. Successful futures researchers work together with funders when defining the general areas of activity and the project objectives. In the process, researchers must take into account the funder's organizational structures, its areas of responsibility, and its decision-making authority. Moreover, they must consider the total spectrum

of persons and stakeholders who are relevant to the project. The decision-making options they recommend must be geared to each actor group they identify (e.g., political decision-makers, member organizations, corporation divisions, etc.).
2. *Apply sound research methods and analysis before identifying decision-making options*:
Futures researchers bear some responsibility for the appropriate implementation of their results. The decision-making options they identify set the range of possible interpretations and paths for implementation. But researchers must also ensure that their results are thorough and accurate; otherwise, they may result in false conclusions or misuse (see "Theoretical Foundation," "Choice and Combination of Methods," "Producing Quality Research," and "Project and Process Management"). In this regard, it is also important that they pitch the results specifically to their research audience (see "Transferability and Communication of Results").
3. *Orient research to the decision-making spaces defined at the start of the project*:
Predefined objectives and decision-making spaces reduce the number of possible measures that can be taken. Research that remains oriented to those foci is more efficient, and more likely to achieve the necessary depth and level of detail. Moreover, the intended actors will be more inclined to implement the recommendations (see "Practical Relevance, Usefulness, and Effectiveness").
4. *Be familiar with the relevant decision-making spaces*:
Futures researchers can competently identify decision-making options only if they are familiar with the corresponding decision-making spaces. Checklists, samples, and other structural aids can help support their work.
5. *Point out opportunities and risks*:
When discussing decision-making options, it is important that researchers point out their respective opportunities and risks. The possible risks should concern both the uncertainty regarding the research—be it the data, the analysis, or the projections—and those that may result from implementing a given course of action. Projects for the political sphere must also assess the regulatory impact with regard to unintended side-effects.

   It is important that researchers clearly specify the conditions under which the identified opportunities and risks are likely to occur. When talking about probable outcomes, they should take care to state the underlying assumptions and indicate any reservations.
6. *Formulate concrete decision-making options*:
To ensure the effectiveness of decision-making options, researchers should rely on the criteria laid out by the S.M.A.R.T principle. Options should be

*s*pecific, *m*easurable in terms of implementation, *a*cceptable to the funder and the stakeholders, *r*ealistic in terms of available resources, and *t*imely in terms of implementation. It is the funder's responsibility to define the measures for implementation, to determine the persons responsible for carrying them out, and to set the budget and time frame. For example, if a project identifies the most attractive areas of innovation, the funder must set new priorities for R&D and reallocate budgets.

Common Shortcomings and Pitfalls

a) *Failing to provide decision-making options*:
   Futures researchers may present trends and scenarios but not offer any decision-making options because the task was not included under the project objectives. Even so, funders may expect such options, and their absence may cause quick relegation of the project report to the circular file, despite a successful presentation. For the sake of effectiveness, therefore, researchers should double-check with funders about their expectations.
b) *Presenting decision-making options that do not cohere with the results*:
   Researchers may present concrete decision-making options that do not logically cohere with the results. This can leave the funder confused and can cast doubt on the legitimacy of the results. It is essential that the derivation of the recommendations from the results be clear and easy to understand.
c) *Offering recommendations that are too general and impracticable*:
   Researchers may formulate the decision-making options in such general terms as to be unusable by the funder and to raise doubts about the quality of their work. They could have avoided the problem by familiarizing themselves with the funder's spaces of action or by involving researchers with more experience in the applicable fields.
d) *Failing to align recommendations with the requirements and standards of the client*:
   Researchers may provide decision-making options that do not satisfy the above-mentioned S.M.A.R.T. criteria. It is important, therefore, that at the beginning of a project, futures researchers familiarize themselves with the standards of the funder and strictly comply with them. For example, researchers may fail to quantify trends and scenarios, which can make it difficult for the funder to utilize the results for its markets and business situations. The wrong time frame can also be problematic. If too long, it often erodes

public support and makes result implementation difficult; if too short, the results can often be insignificant and fall short of the funder's expectations. Hence, the time horizon needs clarifying before the project starts. In particular, researchers should check the compatibility of the time frame with funder's existing strategy as well as with any upcoming strategic decisions or events.

e) *Providing recommendations contrary to a funder's wishes*:
Researchers may provide specific, systematically derived decision-making options without these being stipulated in the contract. This can irritate the funders who want to put aside the issue of implementation and think about possible options without the project's influence. This problem could have been avoided by explicitly agreeing at the start of the project that decision-making options are not to be included in the final report.

f) *Failure to coordinate on the scope of recommendations*:
Researchers may identify decision-making options and recommendations for action that the funder rejects because they narrow its leeway in discussions with stakeholders. Here, researchers cross the boundary between futures research (pointing out options for action on the basis of future assessments) and futures consulting (recommending an action and helping to implement it). If recommendations for action were explicitly agreed on, however, then the problem may lie in the lack of coordination regarding specific conditions such as organizational structures, budget restrictions, etc. In consulting projects, the coordination of the recommendations and the subsequent implementation are independent processes and are often covered by separate projects.

Illustrative Example

An international professional association prepared a scenario study with the help of researchers employed by its members firms. The aim of the study was to envision coming changes in the industry and prepare its member institutions for the future. In addition, the study sought to show that planned legislative projects and regulations could have a negative effect on the industry's international competitiveness, with major economic consequences.

Using advanced software and an experienced programmer, researchers developed scenarios that were both concrete and coherent. Most member firms accepted the scenario's assumptions, even though some of the projected developments were worrying. At a meeting with the association president and the project's steering committee, researchers presented the scenarios and recommended measures. The policy suggestions met with great approval. The same cannot be said of the recommendations for the association and its member firms, which responded indignantly, regarding them as encroachments on still-pending decisions.

The reaction indicates that the researchers did not sufficiently consider the concerns of the funders and stakeholders before making the presentation. To everyone's surprise, however, the approach proved provocative and had a major impact, ultimately transforming the entire industry.

From the funder's point of view, the project overshot the mark. Although the project's mission was to identify options for action, the association took umbrage when given concrete recommendations. That the research ultimately shaped the industry is positive, of course. But the same effect could also have been achieved by doing more to involve the funder and the stakeholders.

Further Reading

Heinzelbecker, K. (2005). Futuring in the chemical industry. *Journal of Business Chemistry, 2*(1), 37–53.

Heinzelbecker, K., & Taylor, A. (2005). Collective forethought: A new paradigm in strategy. *Futures Research Quarterly, 21*(3), 7–21.

Micic, P. (2007). *Die fünf Zukunfts-Brillen—Chancen früher erkennen durch praktisches Zukunftsmanagement* (2nd ed.). Gabal.

Fink, A., & Siebe, A. (2006). *Handbuch Zukunftsmanagement – Werkzeuge der strategischen Planung und Früherkennung*. Campus.

Popp, R., & Schüll, E. (Eds.). (2009). *Zukunftsforschung und Zukunftsgestaltung: Beiträge aus Wissenschaft und Praxis*. Springer.

Project and Process Management 18

Hans-Liudger Dienel

Summary

Almost every human activity can be defined as both a project and as a process: picking up a pencil, writing an article, introducing wind power in Germany. Each project and process has its own dynamic and rhythm, a "natural oscillation" that can be identified, picked up, used, and shaped by project and process management. There are many guidelines and directives that address these process dynamics. There is also a shifting dynamic in the literature on the management of projects and processes. In the sprawling world of guidebooks, preferred management styles and ideologies rapidly change or vanish, only to come back again. In the field of futures studies, the term project is regularly used in two quite different senses: First, project refers to a specific research or consulting project. Second, it refers to the larger economic, social, or political project that is to be researched, advised, or coordinated. While both definitions may differ substantially in their details, they are both subject to general principles, which are summarized here as guidelines. Instructions for successful process management often distinguish process management explicitly from project management. Process management is characterized by its openness to outcomes and to the complexity of initial situations and developments compared with project management, which is more goal-oriented. At the same time, all the standards of professional project management fully apply to process management as well (as

---

H.-L. Dienel (✉)
Department for Work, Technology and Participation, Technische Universität Berlin and Nexus Institut for Cooperation Management, Berlin, Germany
e-mail: Hans-Liudger.Dienel@tu-berlin.de; dienel@nexusinstitut.de

summarized in Blanckenburg et al., 2005). This chapter lists guidelines for both process and project management.

Essentials

Futures researchers design, advise, and shape socio-political processes that arise from diverging interests—in economic terms from struggle over scarce resources, and in political terms from struggle over ideas, power, and influence. The outcome of decisions is uncertain. This is typical and constitutive for almost all processes. Frequently, formal decision-making does not play a dominant role within certain institutions. Rather, the process itself has a subversive dynamic that goes beyond the institutional framework. The process can be described as a game in which actors abide by more or less defined rules that are given by the framework of the social and political system, but which they can also change or transgress.

Thus, a project in applied futures studies and consulting often represents such a process in miniature, mirroring larger socio-political processes in its dynamics or floating like a boat on the ocean of broader social processes. The project manager often does not lead the project but rather supports the project director as facilitator. This is usually referred to as project coordination, and is responsible for all tasks that require coordination between project members. External cooperation management differs from internal project management with direct responsibility, overseeing cooperation processes and problems, and supervising or coaching the project.

The recent literature on change management and project governance often uses a different definition of process management than the one described above. For example, in his book *Process Management: Why Project Management Fails in Complex Decision Making Processes,* Hans de Bruijn, the author of several standard works on the subject, describes process management as situational action that dispenses with prior objectives (see de Bruijn et al., 2010). However, other definitions understand process management as strategic action that pursues a long-term objective with staying power. For an understanding of the concept of *strategy,* which is drawn from the military sphere, it is worth taking a look at the classic work *On War* by Carl von Clausewitz. The well-known management consultant Fredmund Malik writes in the epilogue to a new edition of the seminal work that Clausewitz recognized the essence of strategy when he described war as a highly complex, networked system, "characterized by probability and chance;

18 Project and Process Management

it is open, multidimensional, and non-deterministic, and inherently involves missing information, which means that decision-making and action must always take place under conditions of uncertainty" (Malik in v. Clausewitz, 2005, p. 491).

Nevertheless, projects and processes are not free in their design, but path-dependent, and have an inherent pace and dynamic, which should be taken into account by process management. In the 1960s, Bruce Tuckmann described four typical project phases: forming, storming, norming, and performing. Most projects oscillate between these phases, which need to be shaped by process management. Many guidebooks rightly base their recommendations on the typical phases of projects. They include:

1. the preparatory phase (developing and strategically aligning the project idea, selecting partners, establishing contact with funders, writing the project concept);
2. the constitution and planning phase (getting to know one another, clarifying motives, specifying the involvement of partners, designing the organizational structure, distributing functional roles, understanding the problem and the basic terms, setting internal preconditions and taking hierarchies into account, jointly developing the analysis and definition of problems, making goals operational, planning methods, planning quality assurance, and agreeing on binding forms of partnership); and
3. the implementation and conclusion phase (coordinating the process of implementation, detailed flexible planning, recognizing and managing crises, maintaining dialog with the public, securing interim results, establishing knowledge management, evaluating the project, concluding the project and planning follow-up projects).

In addition to a chronological approach, there is also a thematic approach to project and process management.

From the perspective of communication psychology, for example, process management is about:

1. cooperative culture (relating to each other constructively, giving and receiving feedback, addressing difficult issues, learning by balancing, maintaining motivation);
2. power and leadership (breaking the taboo on power issues, controlling micro-politics, balancing cooperation and competition, distinguishing functional roles and group-dynamic roles, assigning internal leadership functions, adapting project control to the conditions of the partnership); and

3. managing conflicts (clarifying basic attitudes to conflict, taking potentials for conflict seriously, dealing with hot and cold forms of conflict, recognizing and preventing conflict-escalating behavior, distinguishing between factual and relational aspects, approaching conflicts in a power- and/or interest-oriented manner, making use of external conflict consulting, systematically analyzing conflicts with external actors).

A fourth, more instrumental approach to process management focuses on the fundamental principles and techniques of facilitation (participant responsibility, role and tasks of the facilitator, visualization, standard techniques for routine group work, supplementary techniques for problem solving).

Guidelines

The project and process management guidelines presented below follow the typical course of a project.

Guidelines for Project Management

1. *Write the project concept or proposal cooperatively, if possible*:
   Key steps in integrating the major stakeholders in the project can be accomplished by jointly writing the project concept. This cooperative formulation of the project or project proposal differs from the usual application practice, in which time pressure places the lead in the hands of those partners who initiated the project and have a primary interest in its funding. However, the entire research process will be guided by the substance and methodology as specified in the application, and for this reason, it is important in the interest of successful integration that each partner individually formulates the subproject-specific research along with the methods, goals, and competencies. The constitution and planning phase must begin with the application rather than with the approval.
2. *Become acquainted with each other*:
   Who are the partners in the project? In addition to identifying the project partners and their different roles, the beginning of the project work should include an exchange about the focal points, positions, and interests of the individual partners. This is an important measure to support integration. Only the exchange and visualization of ideas about focal points, positions, and

interests can enable the development of common goals and strategies and the integration of knowledge and methods. Even if partners have different substantive positions, they should be able to formulate common interests and agree on common goals and strategies. Sometimes it is helpful to conceptualize a constellation map of the consortium.

3. *Clarify problems and basic terms*:
   At the earliest possible stage in the project, the project partners should—if possible—clarify problems and basic terms. Making a list of key terms related to the common project topic in a common glossary and formulating metaphors for them can facilitate shared understanding.

4. *Jointly develop the analysis and definition of the problem*:
   Jointly collecting, grouping, and weighting the problems to be worked on favors collaboration. A jointly formulated problem can then be placed at the centre of a cause-and-effect analysis. By reflecting on the history of the problem, its causes and consequences, the team can progressively specify the core problem within the framework of this analysis. The results of the analysis can be visualized in a cause-effect diagram. This step should make it easier for the project partners to adopt a common problem definition that everyone can agree upon.

5. *Make goals manageable*:
   Following a cooperative problem definition and analysis, the next essential prerequisites for deriving strategies and methods in the project will be a discussion of common goals and the integration of the separate goals of the project partners. This process of integration is directed toward the definition of various types of goals: positive and negative, general and concrete, simple and complex, explicit and implicit, intermediate and final. The partners should create a goal diagram in which they break down broader common goals into more specific individual goals and reconcile them to their respective individual goals associated with the project. The goals should be formulated positively, that is, they state what the partners want to accomplish and not what they want to prevent. Each individual goal should be checked for its compatibility with the common goals; any implicit goals should also be made clear at this point. The goals are then prioritized according to importance and urgency. In keeping with this set of priorities, it is possible to then compile strategies for achieving the goals.

6. *Plan quality assurance:*
   Project participants may already feel overwhelmed by the usual interim and final evaluations, as the effort devoted to such matters is often misaligned with actual benefits. Instead of filling out endless questionnaires, it is better

to use direct discussions whenever possible to share experiences. However, to help integrate content and methods in the process, internal quality assurance in the form of an internal evaluation is useful. A distinction should be made between structural, process, and outcome quality. Structural quality includes aspects such as the organizational form of the project, decision-making responsibilities, and financial and personnel resources. Process quality refers to the management of information and knowledge exchange, appropriate working methods, adherence to agreements or regular assessment. Outcome quality refers to the intended goal or product of the project. Based on the common goals, strategies and methods for the project, the project partners should agree on standards for the quality areas and develop and agree on a method for internal evaluation. The design of internal quality assurance is developed based on the core questions to be asked in the evaluation process: What is to be assessed? Which sources of information and which methods should be used for this purpose? In what form should the results be presented and how should they be used for quality assurance?

7. *Define and save interim results*:
For the ongoing integration of heterogeneous bodies of knowledge generated in a project over a longer period of time, it is advisable to document interim results that have been previously targeted and defined in a milestone plan. As the milestones are defined, the partners should also agree on the form in which they will be summarized. In addition to the creation of project-internal documents, various forms of publication can be chosen to accomplish this goal. Such safeguarding of interim results also offers the project partners the opportunity to check the status of knowledge integration and methods over the course of the project.

8. *Manage knowledge:*
Knowledge management in the project includes the acquisition, identification, development, representation, preservation, and distribution of knowledge. These aspects are always intertwined. Knowledge integration requires knowledge representation in a form that can be understood by all partners. For this purpose, a "project memory" accessible to all partners should be created with the important documents of the project development. It is important to concentrate on the essentials –the "data graveyards" left by past projects are legion.

Process Management Guidelines

Beyond goal-oriented project management, one aim of futures studies is to manage accompanying socio-political processes. For this reason, the most important methods for process management are also presented here, which are all the more important as processes become larger and more complex. These are not Machiavellian notions of how to achieve and maintain power, but pragmatic, experiential, and carefully considered concepts of process management as participatory governance. They provide experience and encourage innovation, strategizing, tactical thinking, perseverance, and—where necessary—a fresh start in the face of egoism and other obstacles.

1. *Use the group process*:
   Humans are social animals and move in groups. As we know from human ethology, human behaviour and identity is strongly mediated by group belonging. Groups develop their own dynamics, but these can be influenced. Harnessing the power of group dynamics and the desire for identity and belonging is a key tool in managing socio-political processes. This includes the creation of collective partnerships, the formation of internal networks, and the use of communication spaces.
2. *Harness conflict as a catalyst*:
   From a process design perspective, social and political conflicts are not problems to be solved but tools or catalysts for achieving desired goals. It is often useful or strategically smart to escalate conflicts or problems in order to change entrenched structures.
3. *Engage stakeholders*:
   Engaging old, new, or additional stakeholders is a key element in successful process management, especially to resolve conflicts. There is a wide range of possibilities here, including the targeted involvement of a new actor or several new actors; reactivating old actors; initiating open-ended participation; and "embracing" partners while giving them little wiggle room for refusal.
4. *Establish legitimacy*:
   The renowned German sociologist Max Weber argued that political legitimacy is always grounded in one of three different forms of authority: traditional authority, charismatic authority, or rational-legal authority. In this connection, project managers would be wise to note the importance of *inclusive decision-making*, as this is one crucial source of the rational-legal authority that is required to legitimize a political process and undertake collective action.

5. *Exploit institutional competition*:
   From the outside, many organizations look like a monolithic block. But a second glance reveals that within organizations, individual departments and employees often compete with each other, and that many decisions can only be explained by this competition. For the design of many processes, it therefore proves to be very advantageous to recognize and utilize institutional competition.
6. *Stage new beginnings*:
   Stalled processes sometimes need an impetus to get going again or to take a new direction. Staging a new beginning is a process management tool to help set a new course. "And there is a magic in every beginning. That protects us and helps us to live," writes Hermann Hesse in his well-known poem *Stufen* [Steps]. Good process management can make deliberate and creative use of this magic.
7. *Set an agenda*:
   Agenda setting goes beyond the deliberate staging of a new beginning: it sets substantive priorities and creates a work program. For this reason, however, it is more difficult to achieve than a new beginning. Beyond just setting the stage, it also requires the actual, authentic mobilization of actors. Therefore, agenda setting includes the mobilization of stakeholder support.
8. *Use power in a targeted and tailored way*:
   Projects and political processes require the mobilization and use of power. Therefore, it is necessary to openly analyze power dynamics when designing processes. In this connection, it is important to consider how power operates from a practical perspective. It was precisely this quality that made Machiavelli's *The Prince* so shocking and revolutionary, as he explored the practical dimensions of maintaining and exercising power, free from religious dogma or hazy moral considerations.
9. *Make use of personalities and leaders*:
   The quality and outcome of a process are highly dependent on the personalities of those who are involved in it. Therefore, a central element for the success of process management is to give the creative potential of the important actors in the process an opportunity to flourish. This can be achieved, for example, by clearly assigning tasks and roles.
10. *Don't give up hope*:
    Max Weber described politics as "slowly drilling through thick planks with passion and a sense of proportion," addressing the need for perseverance and staying power as a response to ever-changing situational demands. Indeed,

without resilience in the face of setbacks, most processes cannot be brought to a successful conclusion.

Common Shortcomings and Pitfalls

The following is a list of eight typical mistakes that stand in the way of successful process and project management in futures research:

A) *No money for project management*:
Devoting resources to professional project and process management is an investment that pays off in almost all projects, because otherwise collaborative problems may tie up an even larger portion of the available resources. Be sure to realize that it is easy to underestimate the costs and time required for project and process management.
B) *No sense of timing*:
The proper use of pace and rhythm—that is, the right time to start and stop—are critical factors in the success of project and process management; misjudging them can bring about rapid failure. Bismarck, a master of political timing, spoke of listening to the "rustle of God's mantle in history," while the management guru Tom Peters emphasizes seizing the "window of opportunity." The right compromise, a balancing act between courageous action and calm waiting, requires experience and is difficult. It can quickly turn into over-committed impulses or missed opportunities. As futurologist Alwin Tofler noted back in the 1960s, "the future always comes too fast and in the wrong order."
C) *No suitable location for developing a common identity*:
Project management has a spatial dimension that can easily be underestimated. A successful project needs a suitable physical (or virtual) location that is accepted by the project partners as a common space. This could be a shared online presence or a conference center where the project partners meet on neutral ground, either on a recurring or one-off basis.
D) *No common language or images in the project*:
Project and process management is responsible for ensuring that project partners can understand each other. A large element of project management is translation and communication work. Images, maps, and constellations are of particular importance here.

E) *No attention given to the interests of partners*:
   Acceptance of project and process management dwindles among project partners if they see their interests disregarded. The neutrality of the "honest broker" (Bismarck) in balancing the interests of the project partners is crucial for ensuring the legitimacy and acceptance of the project manager. When interests cannot be accommodated, it is especially important to acknowledge and respect them.
F) *No sensitivity to the different paces of organizational development*:
   This pitfall involves a failure to recognize the different speeds of individual and organizational change in a project. A project should be viewed as a convoy in which the slowest vehicle sets the pace—or must be left behind.
G) *No interest in facilitation or participatory processes*:
   For many years, project partners had reservations about modern facilitation and participation formats. But expectations have changed. Today, project partners and stakeholders expect project and process facilitation to employ highly professional methods. Smart project managers skillfully incorporate this interest in learning new methods.
H) *No praise for project partners*:
   We know from the field of education that praise has a stronger behavior-altering effect than criticism. Project management that predominantly criticizes makes life difficult for itself and the team. Praise and encouragement are essential needs, both for individuals and organizations.

Illustrative Example

"We don't need a department for futures studies in our company," Frederick Jameson, the CEO of a logistics company with 6,000 employees, told his press officer Kerstin Meier. "Our future should be discussed in all departments and should also guide and incentivize our actions. I think the scenario idea is good, but you don't need a separate department for that." Jameson, who had taken over the company founded by his grandfather 15 years ago, loved clear, brief, and unpretentious language. He continued to refer to his business as a freight forwarding company rather than a logistics company, since more than half of the employees were truck drivers.

Kerstin Meier left the office somewhat dejected but at the same time curious and proud, since she had taken on the task of launching an internal scenario process for the medium-term future of the freight forwarding company, which she was now to organize over the next twelve months without a support staff, and, of course, in close coordination with the boss and the company's other senior management.

Her plan, which had impressed Jameson, was to give every employee an equal chance to participate in the process, whether driver, secretary, or buyer. To implement this lofty goal, Meier decided to use random selection: every 200th employee on the alphabetical list of salary recipients, i.e. a total of 30 colleagues, would be invited to participate in six all-day workshops on behalf of the entire workforce and to develop a vision for the future. While Jameson had initially been opposed to the idea and wanted to ensure that no important voice would be left out, he was eventually persuaded that random selection would be best to ensure that each employee had an equal chance to participate.

Within the company, too, the news was initially received with some disbelief ("You're not serious"), amusement ("Doesn't the old man know any better?"), and even annoyance ("We have more important things to do right now!"). However, those selected felt honored when they received a letter from the CEO asking them to participate as "future representatives" and assist with the development of a vision for the company. All those who were not selected were expected to comment on the project via the company intranet.

It was amazing to see how the role as "future representative" changed the appearance and self-confidence of those selected. Their input resulted in an astonishingly far-reaching, concrete concept. Jameson was deeply impressed. The central ideas developed by the employees included relocating the company headquarters from its current rural setting to the grounds of the regional airport ("I can't believe that something like that would garner support without resistance!"); the hiring of self-employed truck drivers to make them full-fledged employees ("I wonder if we can afford that?"); and introducing an online communications platform to enable improved communication between truck drivers and dispatchers ("a wonderful idea"). "I wouldn't have dared to suggest some of these things myself," Jameson reflected, "but the neutral process was necessary, particularly for the sensitive issues." It was an important lesson for Jameson: the active involvement of stakeholders not only generates novel perspectives, but also mobilizes support for change.

References

von Blanckenburg, C., Böhm, B., Dienel, H.-L., & Legewie, H. (2005). *Leitfaden für interdisziplinäre Forschergruppen: Projekte initiieren—Zusammenarbeit gestalten.* Steiner.

de Bruijn, H., ten Heuvelhof, E., & in 't Veld, R. (2010). *Process management: Why project management fails in complex decision making processes.* Springer.
Von Clausewitz, C. (2005). *Vom Kriege: Mit einem Nachwort von Fredmund Malik.* Insel. [English: Von Clausewitz, C. (2008). *On War.* Princeton University Press.]

Further Reading

Defila, R., Di Giulio, A., & Scheuermann, M. (2006). *Forschungsverbundmanagement: Handbuch für die Gestaltung inter- und transdisziplinärer Projekte. Management transdisziplinärer Forschungsprozesse.* Birkhäuser.
Defila, R., & Di Giulio, A. (2018). What is it good for? Reflecting and systematizing accompanying research to research programs. *GAIA-Ecological Perspectives for Science and Society, 27,* 97–104.
Henseler, C., & Dienel, H.-L. (2016). Maps of the uncertain: A new approach to communicate scientific ignorance. *Innovation: European Journal of Social Science Research, 30,* 121–136. https://doi.org/10.1080/13511610.2016.1235496
Karlstetter, E., Berkenhagen, J., Legewie, H., & Dienel, H. L. (2003). Transorganisationales Wissensmanagement: Das Forschungsinformationssystem des Bundesministeriums für Verkehr, Bauen und Wohnen. *Wirtschaftspsychologie, 5,* 142–146.
Raschke, J., & Tils, R. (2007). *Politische Strategie: Eine Grundlegung.* VS Verlag.
Schophaus, M., Dienel, H. L., & von Braun, C.-F. (2004). Brücken statt Einbahnstraßen: Lösungsorientiertes Kooperationsmanagement für die interdisziplinäre Forschung. *Wissenschaftsmanagement, 10*(2), 16–26.
Ulbrich, H., Wedel, M., & Dienel, H.-L. (Eds.). (2021). *Internal crowdsourcing in companies: Theoretical foundations and practical applications.* Springer.
Von Prittwitz, V. (2007). *Vergleichende Politikanalyse.* UTB. [English: Von Prittwitz, V. (2012). *Multidimensional political analysis.* FU Berlin. https://userpage.fu-berlin.de/vvp/multidimensional_political_analysis_200612.pdf. Accessed 24 June 2021.]

GPSR Compliance
The European Union's (EU) General Product Safety Regulation (GPSR) is a set of rules that requires consumer products to be safe and our obligations to ensure this.

If you have any concerns about our products, you can contact us on

ProductSafety@springernature.com

In case Publisher is established outside the EU, the EU authorized representative is:

Springer Nature Customer Service Center GmbH
Europaplatz 3
69115 Heidelberg, Germany

www.ingramcontent.com/pod-product-compliance
Ingram Content Group UK Ltd.
Pitfield, Milton Keynes, MK11 3LW, UK
UKHW021324180426
11947UKWH00017B/1424